Composites Science and Technology

Series Editor

Mohammad Jawaid, Laboratory of Biocomposite Technology, Universiti Putra Malaysia, INTROP, Serdang, Malaysia

This book series publishes cutting edge research monographs comprehensively covering topics in the field of composite science and technology. The books in this series are edited or authored by top researchers and professional across the globe. The series aims at publishing state-of-the-art research and development in areas including, but not limited to:

- Conventional Composites from natural and synthetic fibers
- Advanced Composites from natural and synthetic fibers
- Chemistry and biology of Composites and Biocomposites
- Fatigue damage modelling of Composites and Biocomposites
- Failure Analysis of Composites and Biocomposites
- Structural Health Monitoring of Composites and Biocomposites
- Durability of Composites and Biocomposites
- Biodegradability of Composites and Biocomposites
- Thermal properties of Composites and Biocomposites
- Flammability of Composites and Biocomposites
- Tribology of Composites and Biocomposites
- Applications of Composites and Biocomposites

Review Process

The proposal for each volume is reviewed by the main editor and/or the advisory board. The chapters in each volume are individually reviewed single blind by expert reviewers (at least two reviews per chapter) and the main editor.

Ethics Statement for this series can be found in the Springer standard guidelines here - https://www.springer.com/us/authors-editors/journal-author/journal-author-helpdesk/before-you-start/before-you-start/1330#c14214

Vinod Kushvaha · M. R. Sanjay ·
Priyanka Madhushri · Suchart Siengchin
Editors

Machine Learning Applied to Composite Materials

Springer

Editors
Vinod Kushvaha
Department of Civil Engineering—Materials and Structures
Indian Institute of Technology Jammu
Jammu, India

Priyanka Madhushri
Proof of Concept and Innovation Group
Stanley Black & Decker (United States)
Atlanta, GA, USA

M. R. Sanjay
Natural Compsoites Research Group Lab
King Mongkut's University of Technology
Bangkok, Thailand

Suchart Siengchin
King Mongkut's University of Technology
Bangkok, Thailand

ISSN 2662-1819 ISSN 2662-1827 (electronic)
Composites Science and Technology
ISBN 978-981-19-6277-6 ISBN 978-981-19-6278-3 (eBook)
https://doi.org/10.1007/978-981-19-6278-3

This Springer imprint is published by the registered company Springer Nature Singapore Pte Ltd.
The registered company address is: 152 Beach Road, #21-01/04 Gateway East, Singapore 189721, Singapore

Contents

Applications of Machine Learning in the Field of Polymer Composites 1
Aanchna Sharma and Vinod Kushvaha

Image Processing and Machine Learning Methods Applied to Additive Manufactured Composites for Defect Detection and Toolpath Reconstruction 19
Guan Lin Chen and Nikhil Gupta

AI/ML for Quantification and Calibration of Property Uncertainty in Composites 45
Emil Pitz and Kishore Pochiraju

Radial Basis Function-Based Uncertain Low-Velocity Impact Behavior Analysis of Functionally Graded Plates 77
P. K. Karsh, R. R. Kumar, Vaishali, and S. Dey

Application of Machine Learning in Determining the Mechanical Properties of Materials 99
Naman Jain, Akarsh Verma, Shigenobu Ogata, M. R. Sanjay, and Suchart Siengchin

Machine Learning Prediction for the Mechanical Properties of Lightweight Composite Materials 115
Lin Feng Ng and Mohd Yazid Yahya

Ballistic Performance of Bi-layer Graphene: Artificial Neural Network Based Molecular Dynamics Simulations 135
Kritesh Kumar Gupta, Lintu Roy, and Sudip Dey

Quantifying the Sensitivity of Input Parameters in an ANN-Based Committee Networks Model for Estimation of Steel Girder Bridge Load-Ratings 149
Fayaz A. Sofi, Irqab Farooq, Javed A. Bhat, and Manzoor A. Tantray

Estimating Axial Load Capacity of Concrete-Filled Double-Skin Steel Tubular Columns of Multiple Shapes Using Genetic Algorithm-Optimized Artificial Neural Networks 167
Fayaz A. Sofi, Hazim Wani, Mohammad Zakir, and Manzoor A. Tantray

Applications of Machine Learning in the Field of Polymer Composites

Aanchna Sharma and Vinod Kushvaha

1 Introduction

Aerospace, marine, automobile and many other consumer industries are demanding for alternate materials which have an amalgamation of high strength to weight ratio, high toughness, thermal insulation and corrosion resistance along with acoustic damping. These days polymer composites have become the most popular and ideal choice for these industries due to the great amalgam of above-mentioned superior properties that they possess [1–6]. Polymer composites are the multiphase materials that are made up of a continous phase more commonly known as the matrix, and a dispersed phase, generally known as the reinforcement. The matrix phase is essentially a thermosetting/thermoplastic polymer and the typical function of a matrix is to strongly hold the reinforcement and uniformly transfer the load to it [7]. There are various kinds of man-made or natural fibers, particles and whiskers that are used in the form of reinforcement to strengthen the polymeric matrix [8, 9]. Some of the examples are kevlar, aramid, carbon, glass, flax, jute, hemp, coir, sisal, alumina, mica, basalt etc. [7]. Figure 1 is an illustration of the broader classification of polymer composites on the basis of the kind of reinforcing agent used.

Factors like the chemical composition, aspect ratio, extent of the reinforcement and how they are added in consort with the composite manufacturing process significantly affect the final mechanical response that these materials possess [2, 3, 10–14]. If the matrix is reinforced with continuous fibers then the fiber orientation and the stacking order become the deciding parameters while in the case of particulate polymer composites, size and shape of the filler particles are important [15–17]. The interface between the matrix and the reinforcement is another aspect that has significant effect on the mechanical behavior of the polymer composites. The extent

A. Sharma · V. Kushvaha (✉)
Department of Civil Engineering, Indian Institute of Technology Jammu, Jammu, Jammu and Kashmir, India
e-mail: vinod.kushvaha@iitjammu.ac.in

V. Kushvaha et al. (eds.), *Machine Learning Applied to Composite Materials*,
Composites Science and Technology, https://doi.org/10.1007/978-981-19-6278-3_1

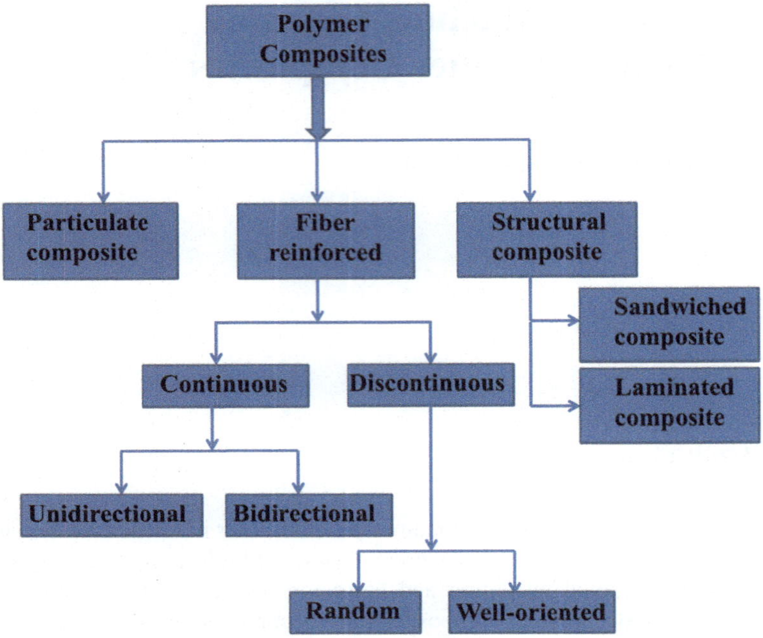

Fig. 1 Classification of polymer composites based on types of reinforcement

of plastic yielding, toughneing, fiber bridging and the overall morphology of such materials get largely affected by the type of interface developed [18, 19]. Another governing factor is the manufacturing process which affects the performance and the load carrying capacity of the end product in the long run [20]. Manufacturing processes like hand lay-up, resin transfer molding, vacuum molding, direct extrusion, pultrusion and compression molding are mostly used and the process choice is made concerning the specific application. Curing temperature, void fraction and the processing speed are additional important parameters that come into play with a particular manufacturing process and lead to the development of residual stresses and strain [21–23]. Therefore, to ramp up the applicability of the polymer composites, it is imperative to study and analyze the combined effect of all these parameters. Selecting the right combination of matrix and reinforcement results in the desirable ultimate response and this tailorability is what gives polymer composites an edge over other conventional materials [24].

Flexible and economical materials suitable for various simultaneous goals are a requirement of most progressive industries. In order to reconnoitre the broader potential of polymer based composites, it is essential to consider all the process parameters together and then optimize the same, this will greatly contribute to the material design and development. However, experimenting on the relevance of each of the regulating parameters is difficult and time-consuming. Also it is extremely difficult to model the complex interactions between the regulating parameters. Over the last few years, material science has progressed from a period of establishing purely computational tools for the disclosure and design of new and complicated materials to the period of creating coupled techniques that improve the quality of results by utilizing the computational forecasts and trial approval. Methods like finite element, molecular dynamic simulations and phase field modelling have been utilized in different fields to demonstrate the material conduct, however the intricacy and the computational severity of these strategies have inspired the researchers to search for different other options [22, 25–31]. In view of this, numerous scientists have turned to machine learning to come up with an optimal design by considering all the important parameters [32, 33]. Machine learning is a subset of artificial intelligence that provides a broader space for effectively exploring and studying the behavior of complex composite materials with limited set of experimental or computational information. Considering the kind of problem to be addressed and the available dataset, there are different machine learning algorithms that are being used in different fields of material science. Though many studies sum up how the machine learning algorithms can be utilized in material science, a large portion of the investigations are restricted in scope by the imperatives of various factors which bring about expanded dimensionality and vulnerability brought about by the uncertainty of the considered information. This chapter aims to summarize the findings of a wide body of research in order to emphasize the wide range potential of machine learning in polymer composites. Henceforth, the chapter is structured as, Sect. 2: an overview of the applications of machine learning in polymer composites, Sect. 3: widely adopted machine learning algorithms, Sect. 4: concluding annotations.

2 An Overview of the Applications of Machine Learning in Polymer Composites

In case of polymer composites, the huge number of existing factors and numerous combinations need to be analysed in order to increase the applications with multi-functional needs. Therefore the researchers are driven to perform more systematic and data-intensive research in the analysis of polymer composites. Combined efforts of experiments and computer simulations have been generating huge chunks of data that need to be analyzed effectively for better insights and understanding. To effectively handle the large and complex data and to draw meaningful inferences, machine learning algorithms are being integrated with material science. Figure 2 highlights the key applications of machine learning algorithms that have been successfully implemented.

In material science, the two mostly used paradigms of machine learning are supervised and unsupervised learning. Supervised learning makes use of a well-labeled dataset and the new predictions are made on the basis of input–output mapping considering the relationship between inputs and the known outputs [35]. While in the later one, the user is allowed to discover the patterns in the given dataset and make

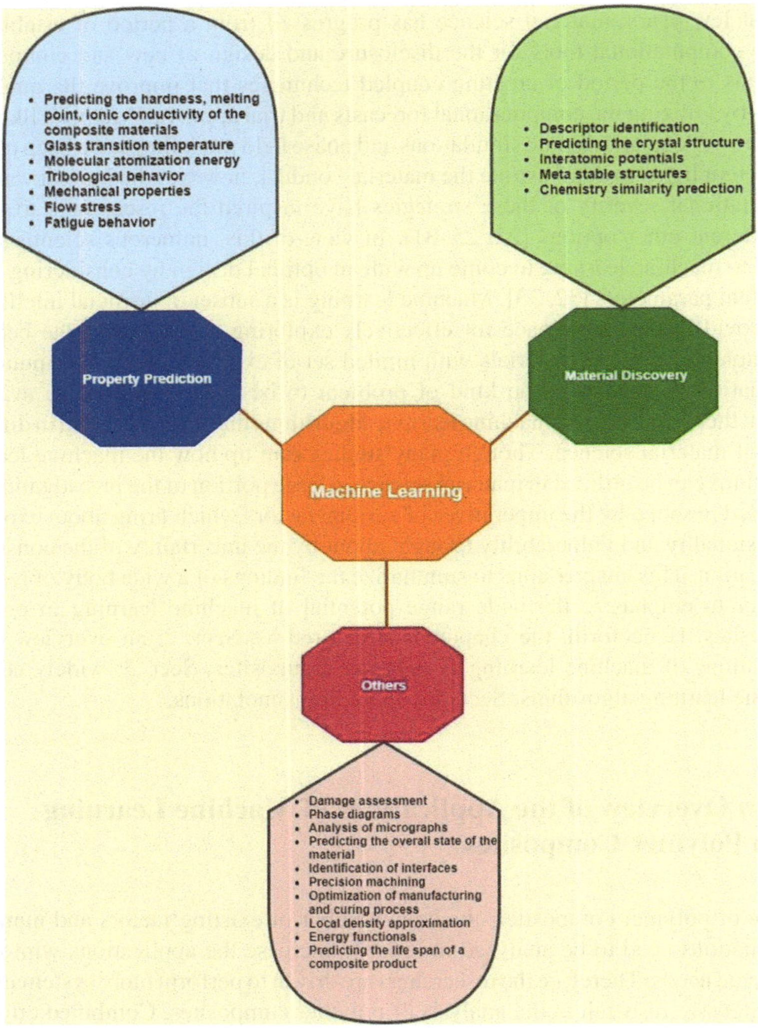

Fig. 2 Applications of machine lerarning in various fields of material science [34]

the predictions accordingly [36, 37]. Supervised learning is more popular when it comes to polymer composites. A typical workflow of implementing the algorithms of supervised learning for making predictions of the material behavior is given in Fig. 3.

In the implementation of any machine learning algorithm, the first step is to prepare the data to be utilized. Raw information consists of huge amount of noise which requires appropriate enhancement in terms of feature engineering that includes reordering, cleaning and then sampling the data. Redundant parameters are deleted in

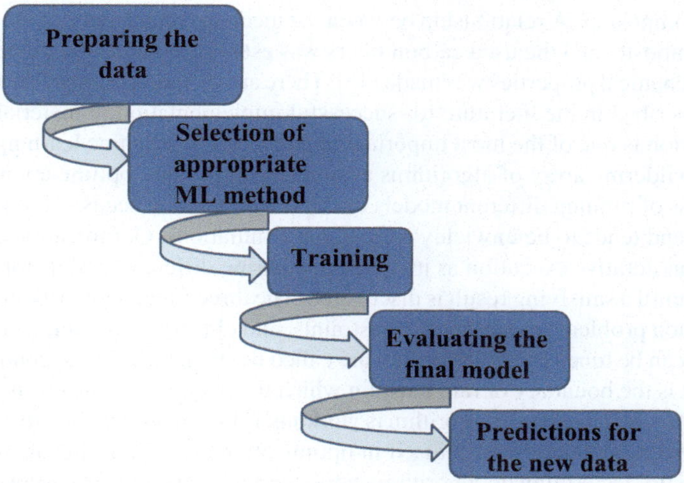

Fig. 3 Workflow of supervised machine learning algorithms

this step so that only the relevant parameters are considered and also the problem of high-dimensionality is avoided. This step helps in improving the learning accuracy of the used algorithm and hence better predictions are made [38, 39].

Once the data is prepared well, a hypothesis function is set that actually performs the mapping between the input and output parameters and also selects the appropriate learning algorithm to be utilized. This algorithm selection greatly depends on the kind of data in hand and also the type of problem which may be regression or classification [40]. The next step involves the training of the model with the preprocessed data. The available data is divided into three categories viz. training, testing and cross-validation dataset. The chosen model learns to map the input and output using the training dataset and for the purpose of hyperparameter tuning, cross-validation dataset is utilized. Testing dataset is used to assess the model's performance on the basis of which predictions are finally made. Different accuracy metrics like mean square error, mean absolute percentage error, coefficient of determination and the confusion matrix are used to evaluate the overall performance of the model.

Machine learning (ML) has been utilized extensively in polymer composites for material property prediction, microstructural characterization, optimization of the process parameters and quantifying the uncertainties caused by the complicated production processes. A group of researchers [41] used density functional theory and a machine learning algorithm to predict properties of long chains of inorganic polymers. Properties like lattice energy, atomization ernergy, spring constant and dielectric constant were predicted and the results from both the approaches were found to be in good agreement. A simple machine learning algorithm, K-nearest neighbor (KNN) was used in a research related to different kinds of bio-nanocomposites in which heat deflection temperature was predicted [42]. Another ML algorithm, decision tree was successfully utilized in modelling the behavior of carbon fiber reinforced polymer

(CFRP) composites. A relationship between the mechanical behavior of the resulting CFRP composite and the used carbon fibers was established and a prediction for the macromechanical properties was made [43]. There are various machine learning algorithms described in the literature for successful implementation in material science. Optimization is one of the most important applications of machine learning because of the bewildering array of algorithms available. Simply put, optimization involves the process of running different models multiple times and hence is computationally intensive and tends to be unwieldy in complex simulations. Optimization algorithm involves an iterative execution as it entails comparing different models for probable solutions until a satisfying result is discovered. The three fundamental elements of an optimization problem are variables, constraints, and objective functions. The parameters that can be tuned by the algorithm are called the variables. The second element, constraint is the boundary or range within which the parameters must be present and the goal for which the entire algorithm is working, is known as the objective function. The two main methodologies utilised in optimization algorithms are stochastic and deterministic. Deterministic algorithms apply certain principles to discover a solution, by ignoring the uncertainty associated with the variable space [44, 45]. However the stochastic approaches model the uncertainties in a probabilistic manner in terms of probability distribution functions [46, 47]. On the basis of design variables, objective function and type of constraints, optimization algorithms can be classified into different categories and the same has been illustrated in Fig. 4.

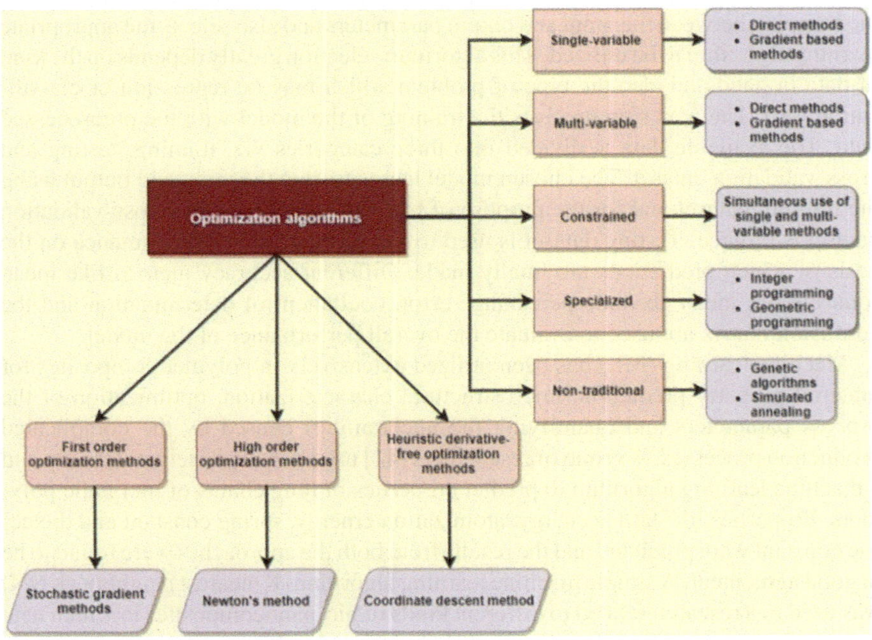

Fig. 4 Classification of optimization algorithms [34]

The approach of machine learning makes it easier to use different optimization algorithms with improved efficiency [48]. To assess and optimize the additive manufacturing processes, ML based approaches have been widely deployed [49]. A research group [50] used a machine learning approch to predict the absorption index of polycarbonate composites reinforced with carbon nano tubes. They used the approach of mutilayer perceptron network and then performed an optimization of the process parameters and consequently the best network was selected for the absorption index prediction. An optimal solution for the was also found by another research group for the critical buckling load and stiffnes of laminates made by using polymer composites [51]. They were able to successfully use the machine learning method as an evolutionary model to reduce the overall computational cost of the optimization process. Many researchers have successfully optimized process parameters and attributes such as the composition of the composite, surface roughness, thrust force, delamination, material properties, and overall structural design using machine learning techniques [50, 52–55].

Because of differences in material properties, measurement uncertainty, test set-up restrictions, operational environment, and imprecise geometrical features, computational evaluations of polymer composites frequently encounter uncertainties [56, 57]. Parametric input uncertainties, initial and boundary conditions, computational and numerical uncertainties resulting from inevitable assumptions and approximations, and the inherent inaccuracy of the model bring a significant deviation of material response from the expected one. Quantifying these uncertainties is essential to ensure the reliability of simulation results and understanding the risks before final product selection. In this view, various machine learning algorithms have been instrumental. Gaussian process, a supervised machine learning algorithm have been used to quantify the inherent uncertainties in the simulation of fiber reinforced composites [58]. The framework of Bayesian inference has also been successfully utilized for the purpose of uncertainty quantification of electrical conductivity of the carbon nanotube reinforced polymer composites [59].

Aerospace, automotive, marine, and many other highly advanced industries are increasingly turning to polymer composites for product design where performance reliability is a key factor. It is critical to assess the dependability of complex systems, including the detection of abnormalities, for long-term development, robust design, and safe operation [60]. Given the many structural and non-structural applications of polymer composites and the different types of stress these composites experience during their useful life, a reliability study for their safe application is essential. Recently, researchers have started exploiting machine learning-based methods to assess the reliability of complex systems in a computationally efficient manner [61, 62]. Algorithms like artificial neural network, support vector machine, random forest and genetic algorithms have been reported in literature for determining the reliability index successfully in the context of polymer composites [63–66].

3 Generally Used Machine Learning Algorithms in Polymer Composites

Researchers are constantly investigating the exploitation of the growing capabilities of machine learning algorithms in the quest of making advancements in polymer composites by discovering new materials, tailoring application oriented properties of composite materials, and evaluating the various affecting parameters in order to come up with an optimized yet feasible design. Such efforts have yielded a number of effective outcomes, which are discussed in the following subsections.

3.1 Neural Networks

Neural network is a mathematical tool that imitates the biological nervous system to solve a broad spectrum of scientific and engineering problems. Akin to the human brain, there are a number of processors (neurons) that work parallely and process the information that is passed over the layers. The first layer commonly known as the input layer, receives all the preprocessed data. There is an intermediate layer also known as the hidden layer that makes all the mathematical computations. Then finally, there is an output layer which gives the final predictions. This kind of network is known as artificial neural network (ANN). The efficiency and robustness of this method lies in its ability to use nonlinear mapping functions to process hefty amounts of data with large covariate spaces. Predictor functions like multilayer perceptron and radial basis function are often used and their purpose is to reduce the prediction error.

ANN is one of the most popular machine learning algorithms for data categorization, prediction, clustering, pattern recognition, and image identification [67–70]. This algorithm has been successfully used by many researchers to predict the mechanical properties of polymer based composites that too with limited experimentation [71–73]. For exploring the plaussible applications of carbon nanotube based polymer composites in damage detecting sensors, the methodology of ANN was used [74]. One of the important observations during this study was that each simulation run using finite element method took 3.5 h to complete, wheras ANN produced almost same results in less than 0.2 s. In order to anticipate the tribological properties of polymer composites, neural networks have been proposed as a new technique [75]. Literature suggests that neural networks have been efficaciously implemented by the research community for the purpose of prediction and optimization of the load-carrying capability, wear rate, delamination, fracature toughness and acoustic properties of different types of polymer composites [67, 76–80]. A research group [81] used the framework of artificial neural network to predict the tribological and mechanical behavior of glass-filled epoxy composites by considering the composition and test parameters as the input functions. This ANN methodology has also been for the purpose of locating defects in the interfacial zone of polymer composites [82].

Some neural networks are based on fuzzy inference system and are known as adaptive neuro fuzzy networks. It is also characterised as a universal estimator because it combines the advantages of neural networks and fuzzy logics into a single architecture. Adaptive neural networks have been implemented to predict the overall mechanical response of glass reinforced polymer composites when subjected to impact load [83]. Polynomial neural network (PNN) is another class of neural networks that has lately gained interest in polymer composites for material behvior prediction. Polynomial neural networks-based predictive modelling has been demonstrated to be more accurate than other fuzzy models [84]. When the output function is a higher order polynomial, PNNs are the best choice. Convolutional neural network is another important class of neural networks that is mostly used for analyzing image-based data [85]. Jiang and his research group used material composition, temperature, normal load and the sliding speed as input parameters and used the algorithm of ANN to predict the wear characteristics along with the mechanical properties of short fiber reinforced polyamide composites [81]. In another study, adaptive neural networks were used to predict the mechanical response of glass-filled polymer composites when fiber content, feed rate and the mixing temperature were used as the input parameters [83]. The framework of CNN was used to predict the strength and stiffness of polymer composites by using the microstructural images of the composite in the input space [85].

3.2 Logistic Regression

Logistic regression, often known as the classification algorithm, is commonly used for predicting discrete values. This supervised learning approach involves applying activation functions and adding a threshold value to convert a linear equation to a logistic one, as well as mapping the existing values between 0 and 1. Fitting a curve to the available data yields probabilistic predictions in this scenario [86]. Logistic regression makes use of a sigmoid function as follows:

$$y = \frac{1}{(1 + e^{-x})} \tag{1}$$

Owing to the simplicity of this algorithm, it has been extensively used by many for the purpose of classification and prediction [87, 88]. Logistic regression was used by a group of researchers to explain the correlation between the energy density and polymerization force [89]. Another group implemented the framework of logistic regression to predict the overall mechanical behavior of various polymer composites made by different kinds of matrix and reinforcing agents [90]. The same algorithm has been successfully applied to adhesive joints made of fiber reinforced polymer composites in order to identify different modes of damage/failure in it [91]. Logistic regression has applications in assessing the polymeric stability at different temperature ranges, classifying defects and predicting the electrical conductivity along with the thermal response of the resulting composite [86, 92, 93].

3.3 Gaussian Process

The Gaussian process (GP) is a stochastic non-parametric process that can be used to solve non-linear problems [94, 95]. Gaussian process is one of the Naive Bayer's versions that is being utilised to bridge the gap between computer simulations and physical situations. Simple parametric approaches sometimes fall short of accuracy and efficacy when datasets get more complicated. In such cases, neural network implementation becomes difficult. However, the advent of kernel based method, gaussian process has made it possible to draw unbiased and accurate inferences with a small set of training data. In essence, a Gaussian process is a collection of random variables with the premise that all the variables (predictor and target) have the same Gaussian distribution.

A research group used piezoelectric parameters as predictor variables and acoustic parameters as target and used gaussian process to predict the remaining useful life and the damage states of polymer composites [96]. Another group used the B-scans of the composite materials after the material being subjected to delamination at different sites and characterized the impact failure using the framework of gaussian process [97]. Another major observation was the reduction in the computational time by 95%. Interfacial properties of polymer composites which are anyway difficult to be investigated were explored by means of correlating the feature space with the dielectric and viscoelastic properties in a GP framework [98]. The algorithm of Gaussian process has been shown to be quite efficient and robust specially in the case of smaller datsets with high degree of non-linearity.

3.4 Support Vector Machine

It is another kernel-based machine learning algorithm that falls into the category of supervised learning. Support vector machine (SVM) can be applied to both, regression as well as classification problems. SVM has long been considered the most classic method in the broad domain of machine learning when it comes to addressing classification issues involving large amounts of data. SVM works on the principle of optimization, in which the hyperplane is set so that the distance between data points also known as support vectors, and the data separator is as large as possible. Support vector machines have a wide range of applications because they employ kernelization, which allows the algorithm to simulate non-linear decision boundaries by adding an extra dimension to the input [99–101].

In the investigation of polymer composites, SVM-based machine learning models have been widely used. In the instance of CFRP composites, a group of researchers employed support vector regression to forecast the position of impact and measure the energy of the entire event [102]. A structure property relationship model was developed using the framework of supprt vector machine that facilitated in predicting the mechanical properties of polymer nanocomposites corresponding to the variations made in the filler content and the processing temperature [103]. The same algorithm was successfully used in evaluating the bond strength of fiber reinforced polymer with concrete [104]. Also, support vector regression based prediction and optimization model was used to design conductive carbon fiber reinforced polymer composites [105]. SVM has shown promising results when it comes to the classification of matrix cracking, damage detection and quantification and risk factor assessment [106–108].

4 Conclusion and the Future Scope

This chapter summarizes the current developments in the applications of various machine learning techniques to polymer-based composites. Different kinds of analyses that can be performed using machine learning (like prediction, optimization, uncertainty quantification and reliability assessment) are discussed. An overview of commonly used algorithms and their implementation to predict and optimize the material behavior is presented. The progress made in the area of machine learning indicates that the material database can be effectively used to identify trends, patterns and mathematical relations for discovering new materials. Appropriate techniques can be used to identify salient features of different materials under different conditions which can help coming up with an optimum material designing approach.

References

1. van den Ende DA, van de Wiel HJ, Groen WA, van der Zwaag S (2011) Direct strain energy harvesting in automobile tires using piezoelectric PZT–polymer composites. Smart Mater Struct 21:015011. https://doi.org/10.1088/0964-1726/21/1/015011
2. Kushvaha V, Branch A, Tippur H (2014) Effect of loading rate on dynamic fracture behavior of glass and carbon fiber modified epoxy. In: Song B, Casem D, Kimberley J (eds) Dynamic behavior of materials, vol 1. Springer International Publishing, Cham, pp 169–176
3. Kushvaha V, Tippur H (2014) Effect of filler shape, volume fraction and loading rate on dynamic fracture behavior of glass-filled epoxy. Compos B Eng 64:126–137. https://doi.org/10.1016/j.compositesb.2014.04.016
4. Marsh G (2006) Composites get in deep with new-generation engine. Reinf Plast 50:26–29. https://doi.org/10.1016/S0034-3617(06)71188-2
5. Sanjay MR, Arpitha GR, Naik LL et al (2016) Applications of natural fibers and its composites: an overview. Nat Resour 7:108–114. https://doi.org/10.4236/nr.2016.73011
6. Sharma A, Khan VC, Balaganesan G, Kushvaha V (2020) Performance of nano-filler reinforced composite overwrap system to repair damaged pipelines subjected to quasi-static and impact loading. J Fail Anal Prev 20:2017–2028. https://doi.org/10.1007/s11668-020-01013-6

7. Rajak DK, Pagar DD, Kumar R, Pruncu CI (2019) Recent progress of reinforcement materials: a comprehensive overview of composite materials. J Market Res 8:6354–6374. https://doi.org/10.1016/j.jmrt.2019.09.068
8. Barbero EJ (2017) Introduction to composite materials design. CRC Press, Boca Raton
9. Qin QH (2015) 1—introduction to the composite and its toughening mechanisms. In: Qin Q, Ye J (eds) Toughening mechanisms in composite materials. Woodhead Publishing, UK, pp 1–32
10. Afrouzian A, Movahhedi Aleni H, Liaghat G, Ahmadi H (2017) Effect of nano-particles on the tensile, flexural and perforation properties of the glass/epoxy composites. J Reinf Plast Compos 36:900–916. https://doi.org/10.1177/0731684417694753
11. Bharath KN, Madhushri P, Gowda TGY et al (2020) Alkaline effect on characterization of discarded waste of moringa oleifera fiber as a potential eco-friendly reinforcement for biocomposites. J Polym Environ. https://doi.org/10.1007/s10924-020-01818-4
12. Esnaola A, Tena I, Saenz-Dominguez I et al (2018) Effect of the manufacturing process on the energy absorption capability of GFRP crush structures. Compos Struct 187:316–324. https://doi.org/10.1016/j.compstruct.2017.12.079
13. Kathi J, Rhee K-Y, Lee JH (2009) Effect of chemical functionalization of multi-walled carbon nanotubes with 3-aminopropyltriethoxysilane on mechanical and morphological properties of epoxy nanocomposites. Compos A Appl Sci Manuf 40:800–809. https://doi.org/10.1016/j.compositesa.2009.04.001
14. Kushvaha V, Tippur H (2013) Effect of filler particle shape on dynamic fracture behavior of glass-filled epoxy. In: Chalivendra V, Song B, Casem D (eds) Dynamic behavior of materials, vol 1. Springer, New York, pp 513–522
15. Kanda M, Puggal S, Dhall N, Sharma A (2018) Recent developments in the fabrication, characterization, and properties enhancement of polymer nanocomposites: a critical review. Mater Today Proc 5:28243–28252
16. Kushvaha V (2016) Synthesis, processing and dynamic fracture behavior of particulate epoxy composites with conventional and hierarchical micro-/nano-fillers
17. Osman MA, Mittal V, Lusti HR (2004) The aspect ratio and gas permeation in polymer-layered silicate nanocomposites. Macromol Rapid Commun 25:1145–1149. https://doi.org/10.1002/marc.200400112
18. Jesson DA, Watts JF (2012) The interface and interphase in polymer matrix composites: effect on mechanical properties and methods for identification. Polym Rev 52:321–354. https://doi.org/10.1080/15583724.2012.710288
19. Reifsnider KL (1994) Modelling of the interphase in polymer-matrix composite material systems. Composites 25:461–469. https://doi.org/10.1016/0010-4361(94)90170-8
20. Srivastava S, Sharma A, Kushvaha V (2021) Applications of additive manufacturing. In: Mavinkere Rangappa S, Gupta MK, Siengchin S, Song Q (eds) Additive and subtractive manufacturing of composites. Springer, Singapore, pp 201–226
21. McCartney LN, Kelly A (1989) New theoretical model of stress transfer between fibre and matrix in a uniaxially fibre-reinforced composite. Proc R Soc Lond A Math Phys Sci 425:215–244. https://doi.org/10.1098/rspa.1989.0104
22. Meijer G, Ellyin F, Xia Z (2000) Aspects of residual thermal stress/strain in particle reinforced metal matrix composites. Compos B Eng 31:29–37. https://doi.org/10.1016/S1359-8368(99)00060-8
23. Tyson WR, Davies GJ (1965) A photoelastic study of the shear stresses associated with the transfer of stress during fibre reinforcement. Br J Appl Phys 16:199–205. https://doi.org/10.1088/0508-3443/16/2/313
24. Kushvaha V, Anandkumar S, Madhushri P (2019) Dynamic fracture toughness index: a new integrated methodology for mode-I fracture behaviour of polymer composite under impact loading. Mater Res Express. https://doi.org/10.1088/2053-1591/ab4e35
25. Frankland SJV, Harik VM, Odegard GM et al (2003) The stress–strain behavior of polymer–nanotube composites from molecular dynamics simulation. Compos Sci Technol 63:1655–1661. https://doi.org/10.1016/S0266-3538(03)00059-9

26. Garg A, Hasan A, Maloney CE (2020) Mesoscale analysis of homogeneous dislocation nucleation. J Appl Mech 86. https://doi.org/10.1115/1.4043885
27. Hemath M, Rangappa SM, Kushvaha V et al (2020) A comprehensive review on mechanical, electromagnetic radiation shielding, and thermal conductivity of fibers/inorganic fillers reinforced hybrid polymer composites. Polym Compos. http://doi.org/10.1002/pc.25703
28. Li Y, Wang S, Wang Q, Xing M (2018) Enhancement of fracture properties of polymer composites reinforced by carbon nanotubes: a molecular dynamics study. Carbon 129:504–509. https://doi.org/10.1016/j.carbon.2017.12.029
29. Liu Z, Zhang L, Poyraz S et al (2014) An ultrafast microwave approach towards multicomponent and multi-dimensional nanomaterials. RSC Adv 4:9308–9313. https://doi.org/10.1039/C3RA47086E
30. Rossman T, Kushvaha V, Dragomir-Daescu D (2016) QCT/FEA predictions of femoral stiffness are strongly affected by boundary condition modeling. Comput Methods Biomech Biomed Eng 19:208–216. https://doi.org/10.1080/10255842.2015.1006209
31. Talebi H, Silani M, Bordas SPA et al (2014) A computational library for multiscale modeling of material failure. Comput Mech 53:1047–1071. https://doi.org/10.1007/s00466-013-0948-2
32. Sharma A, Madhushri P, Kushvaha V, Subramaniyan AK (2020) Prediction of the fracture toughness of silicafilled epoxy composites using K-nearest neighbor (KNN) method. In: 2020 international conference on computational performance evaluation (ComPE), Shillong, India, pp 194–198. http://doi.org/10.1109/ComPE49325.2020.9200093
33. Wani I, Sharma A, Kushvaha V et al (2020) Effect of pH, volatile content, and pyrolysis conditions on surface area and O/C and H/C ratios of biochar: towards understanding performance of biochar using simplified approach. J Hazard Toxic Radioactive Waste 24:04020048. https://doi.org/10.1061/(ASCE)HZ.2153-5515.0000545
34. Sharma A, Mukhopadhyay T, Rangappa SM et al (2022) Advances in computational intelligence of polymer composite materials: machine learning assisted modeling, analysis and design. Arch Computat Methods Eng. http://doi.org/10.1007/s11831-021-09700-9
35. Riedmiller M (1994) Advanced supervised learning in multi-layer perceptrons—from backpropagation to adaptive learning algorithms. Comput Stand Interfaces 16:265–278. https://doi.org/10.1016/0920-5489(94)90017-5
36. Barlow HB (1989) Unsupervised learning. Neural Comput 1:295–311. https://doi.org/10.1162/neco.1989.1.3.295
37. Hand DJ (2007) Principles of data mining. Drug-Saf 30:621–622. https://doi.org/10.2165/00002018-200730070-00010
38. Blum AL, Langley P (1997) Selection of relevant features and examples in machine learning. Artif Intell 97:245–271. https://doi.org/10.1016/S0004-3702(97)00063-5
39. Yu L, Liu H (2004) Efficient feature selection via analysis of relevance and redundancy. J Mach Learn Res 5:1205–1224
40. Sleeman D, Rissakis M, Craw S et al (1995) Consultant-2: pre- and post-processing of machine learning applications. Int J Hum Comput Stud 43:43–63. https://doi.org/10.1006/ijhc.1995.1035
41. Pilania G, Wang C, Jiang X et al (2013) Accelerating materials property predictions using machine learning. Sci Rep 3:2810. https://doi.org/10.1038/srep02810
42. Daghigh V, Lacy TE, Daghigh H et al (2020) Heat deflection temperatures of bio-nano-composites using experiments and machine learning predictions. Mater Today Commun 22:100789. https://doi.org/10.1016/j.mtcomm.2019.100789
43. Qi Z, Zhang N, Liu Y, Chen W (2019) Prediction of mechanical properties of carbon fiber based on cross-scale FEM and machine learning. Compos Struct 212:199–206. https://doi.org/10.1016/j.compstruct.2019.01.042
44. Francisco M, Revollar S, Vega P, Lamanna R (2005) A comparative study of deterministic and stochastic optimization methods for integrated design of processes. IFAC Proc 38:335–340. https://doi.org/10.3182/20050703-6-CZ-1902.00917
45. Sun S (2013) A review of deterministic approximate inference techniques for Bayesian machine learning. Neural Comput Appl 23:2039–2050. https://doi.org/10.1007/s00521-013-1445-4

46. Pedro HTC, Coimbra CFM, David M, Lauret P (2018) Assessment of machine learning techniques for deterministic and probabilistic intra-hour solar forecasts. Renew Energy 123:191–203. https://doi.org/10.1016/j.renene.2018.02.006
47. Sharma A, Mukhopadhyay T, Kushvaha V (2022) Experimental data-driven uncertainty quantification for the dynamic fracture toughness of particulate polymer composites. Eng Fract Mech 273:108724. https://doi.org/10.1016/j.engfracmech.2022.108724
48. Kalita K, Mukhopadhyay T, Dey P, Haldar S (2020) Genetic programming-assisted multi-scale optimization for multi-objective dynamic performance of laminated composites: the advantage of more elementary-level analyses. Neural Comput Appl 32:7969–7993. https://doi.org/10.1007/s00521-019-04280-z
49. Baturynska I, Semeniuta O, Martinsen K (2018) Optimization of process parameters for powder bed fusion additive manufacturing by combination of machine learning and finite element method: a conceptual framework. Procedia CIRP 67:227–232. https://doi.org/10.1016/j.procir.2017.12.204
50. Sidi Salah L, Chouai M, Danlée Y et al (2020) Simulation and optimization of electromagnetic absorption of polycarbonate/CNT composites using machine learning. Micromachines 11:778. https://doi.org/10.3390/mi11080778
51. Nik MA (2012) Surrogate-based multi-objective optimization of a composite laminate with curvilinear fibers. Compos Struct 8
52. Arian Nik M, Fayazbakhsh K, Pasini D, Lessard L (2012) Surrogate-based multi-objective optimization of a composite laminate with curvilinear fibers. Compos Struct 94:2306–2313. https://doi.org/10.1016/j.compstruct.2012.03.021
53. Duan K, He Y, Li Y et al (2019) Machine-learning assisted coarse-grained model for epoxies over wide ranges of temperatures and cross-linking degrees. Mater Des 183:108130. https://doi.org/10.1016/j.matdes.2019.108130
54. Marín L, Trias D, Badalló P et al (2012) Optimization of composite stiffened panels under mechanical and hygrothermal loads using neural networks and genetic algorithms. Compos Struct 94:3321–3326. https://doi.org/10.1016/j.compstruct.2012.04.024
55. Sarkar S, Vinay S, Raj R et al (2019) Application of optimized machine learning techniques for prediction of occupational accidents. Comput Oper Res 106:210–224. https://doi.org/10.1016/j.cor.2018.02.021
56. Mukhopadhyay T, Chakraborty S, Dey S et al (2017) A critical assessment of kriging model variants for high-fidelity uncertainty quantification in dynamics of composite shells. Arch Computat Methods Eng 24:495–518. https://doi.org/10.1007/s11831-016-9178-z
57. Naskar S, Mukhopadhyay T, Sriramula S (2019) Spatially varying fuzzy multi-scale uncertainty propagation in unidirectional fibre reinforced composites. Compos Struct 209:940–967. https://doi.org/10.1016/j.compstruct.2018.09.090
58. Bostanabad R, Liang B, Gao J et al (2018) Uncertainty quantification in multiscale simulation of woven fiber composites. Comput Methods Appl Mech Eng 338:506–532. https://doi.org/10.1016/j.cma.2018.04.024
59. Doh J, Park S-I, Yang Q, Raghavan N (2021) Uncertainty quantification of percolating electrical conductance for wavy carbon nanotube-filled polymer nanocomposites using Bayesian inference. Carbon 172:308–323. https://doi.org/10.1016/j.carbon.2020.09.092
60. Naresh K, Shankar K, Velmurugan R (2018) Reliability analysis of tensile strengths using Weibull distribution in glass/epoxy and carbon/epoxy composites. Compos B Eng 133:129–144. https://doi.org/10.1016/j.compositesb.2017.09.002
61. Behnia A, Ranjbar N, Chai HK, Masaeli M (2016) Failure prediction and reliability analysis of ferrocement composite structures by incorporating machine learning into acoustic emission monitoring technique. Constr Build Mater 122:823–832. https://doi.org/10.1016/j.conbuildmat.2016.06.130
62. Shuang LH, Zhou LZ, Feng Y (2006) Support vector machines for structural reliability analysis. Appl Math Mech. https://doi.org/10.1007/s10483-006
63. Ahmad A, Khan Q, Uz Z, Raza A (2020) Reliability analysis of strength models for CFRP-confined concrete cylinders. Compos Struct 244:112312. https://doi.org/10.1016/j.compstruct.2020.112312

64. Alsina EF, Chica M, Trawiński K, Alberto R (2018) On the use of machine learning methods to predict component reliability from data-driven industrial case studies. Int J Adv Manuf Technol 94:2419–2433. https://doi.org/10.1007/s00170-017-1039-x
65. Tan X, Bi W, Hou X, Wang W (2011) Reliability analysis using radial basis function networks and support vector machines. Comput Geotech 38:178–186. https://doi.org/10.1016/j.compgeo.2010.11.002
66. Yan F, Lin Z (2016) New strategy for anchorage reliability assessment of GFRP bars to concrete using hybrid artificial neural network with genetic algorithm. Compos B Eng 92:420–433. https://doi.org/10.1016/j.compositesb.2016.02.008
67. Argatov I (2019) Artificial neural networks (ANNs) as a novel modeling technique in tribology. Front Mech Eng 5. https://doi.org/10.3389/fmech.2019.00030
68. Kushvaha V, Sharma A (2021) Dimensional analysis for predicting the fracture behavior of particulate polymer composite under the effect of impact loading. In: Fracture failure analysis of fiber reinforced polymer matrix composites. Springer, Singapore
69. Sharma A, Anand Kumar S, Kushvaha V (2020) Effect of aspect ratio on dynamic fracture toughness of particulate polymer composite using artificial neural network. Eng Fract Mech 228:106907. https://doi.org/10.1016/j.engfracmech.2020.106907
70. Sharma A, Kushvaha V (2020) Predictive modelling of fracture behaviour in silica-filled polymer composite subjected to impact with varying loading rates using artificial neural network. Eng Fract Mech 239:107328. https://doi.org/10.1016/j.engfracmech.2020.107328
71. Mucha W, Kuś W, Viana JC, Nunes JP (2020) Operational load monitoring of a composite panel using artificial neural networks. Sensors 20:2534. https://doi.org/10.3390/s20092534
72. Nayak SK, Satapathy A, Mantry S (2020) Processing and wear response study of glass-polyester composites with waste marble dust as particulate filler. Polym Compos 41:2263–2273. https://doi.org/10.1002/pc.25537
73. Sibi P, Jones SA, Siddarth P (2005) Analysis of different activation functions using back propagation neural networks. J Theor Appl Inf Technol 47:5
74. Matos MAS, Pinho ST, Tagarielli VL (2019) Application of machine learning to predict the multiaxial strain-sensing response of CNT-polymer composites. Carbon 146:265–275. https://doi.org/10.1016/j.carbon.2019.02.001
75. Rout A, Satapathy A (2012) Analysis of dry sliding wear behaviour of rice husk filled epoxy composites using design of experiment and ANN. Procedia Eng 38:1218–1232. https://doi.org/10.1016/j.proeng.2012.06.153
76. Garg A, Wani I, Zhu H, Kushvaha V (2022) Exploring efficiency of biochar in enhancing water retention in soils with varying grain size distributions using ANN technique. Acta Geotech 17:1315–1326. https://doi.org/10.1007/s11440-021-01411-6
77. Hayajneh M, Hassan AM, Alrashdan A, Mayyas AT (2009) Prediction of tribological behavior of aluminum–copper based composite using artificial neural network. J Alloy Compd 470:584–588. https://doi.org/10.1016/j.jallcom.2008.03.035
78. Negro C, Alonso A, Blanco A, Tijero J (2005) Breaking load and bending strength prediction in manufacture of fibre cement composites using artificial neural networks and a flocculation sensor. Compos A Appl Sci Manuf 36:1617–1626. https://doi.org/10.1016/j.compositesa.2005.04.008
79. Shuvho BA, Chowdhury MA, Debnath UK (2019) Analysis of artificial neural network for predicting erosive wear of nylon-12 polymer. MPC 8:288–300. https://doi.org/10.1520/MPC20180164
80. Yeh I-C (1998) Modeling of strength of high-performance concrete using artificial neural networks. Cem Concr Res 28:1797–1808. https://doi.org/10.1016/S0008-8846(98)00165-3
81. Jiang Z, Zhang Z, Friedrich K (2007) Prediction on wear properties of polymer composites with artificial neural networks. Compos Sci Technol 67:168–176. https://doi.org/10.1016/j.compscitech.2006.07.026
82. Farhangdoust S, Tashakori S, Baghalian A et al (2019) Prediction of damage location in composite plates using artificial neural network modeling. In: Sensors and smart structures technologies for civil, mechanical, and aerospace systems 2019. International Society for Optics and Photonics, p 109700I

83. Fazilat H, Ghatarband M, Mazinani S et al (2012) Predicting the mechanical properties of glass fiber reinforced polymers via artificial neural network and adaptive neuro-fuzzy inference system. Comput Mater Sci 58:31–37. https://doi.org/10.1016/j.commatsci.2012.01.012
84. Oh S-K, Pedrycz W, Park B-J (2003) Polynomial neural networks architecture: analysis and design. Comput Electr Eng 29:703–725. https://doi.org/10.1016/S0045-7906(02)00045-9
85. Abueidda DW, Almasri M, Ammourah R et al (2019) Prediction and optimization of mechanical properties of composites using convolutional neural networks. Compos Struct 227:111264. https://doi.org/10.1016/j.compstruct.2019.111264
86. Cao R, Naya S, Artiaga R et al (2004) Logistic approach to polymer degradation in dynamic TGA. Polym Degrad Stab 85:667–674. https://doi.org/10.1016/j.polymdegradstab.2004.03.006
87. Mallela UK, Upadhyay A (2016) Buckling load prediction of laminated composite stiffened panels subjected to in-plane shear using artificial neural networks. Thin-Walled Struct 102:158–164. https://doi.org/10.1016/j.tws.2016.01.025
88. Zhu C, Idemudia CU, Feng W (2019) Improved logistic regression model for diabetes prediction by integrating PCA and K-means techniques. Inf Med Unlocked 17:100179. https://doi.org/10.1016/j.imu.2019.100179
89. Sakaguchi RL, Wiltbank BD, Murchison CF (2004) Contraction force rate of polymer composites is linearly correlated with irradiance. Dent Mater 20:402–407. https://doi.org/10.1016/j.dental.2003.11.004
90. Gu GX, Chen C-T, Richmond DJ, Buehler MJ (2018) Bioinspired hierarchical composite design using machine learning: simulation, additive manufacturing, and experiment. Mater Horiz 5:939–945. https://doi.org/10.1039/C8MH00653A
91. Xu H, Liu R, Choudhary A, Chen W (2015) A machine learning-based design representation method for designing heterogeneous microstructures. J Mech Des 137. https://doi.org/10.1115/1.4029768
92. Berge TLL, Lygre GB, Lie SA, Björkman L (2018) Polymer-based dental filling materials placed during pregnancy and risk to the foetus. BMC Oral Health 18:144. https://doi.org/10.1186/s12903-018-0608-1
93. Osburg V-S, Strack M, Toporowski W (2016) Consumer acceptance of wood-polymer composites: a conjoint analytical approach with a focus on innovative and environmentally concerned consumers. J Clean Prod 110:180–190. https://doi.org/10.1016/j.jclepro.2015.04.086
94. Rasmussen CE (2004) Gaussian processes in machine learning. In: Bousquet O, von Luxburg U, Rätsch G (eds) Advanced lectures on machine learning: ML summer schools 2003, Canberra, Australia, February 2–14, 2003, Tübingen, Germany, August 4–16, 2003, revised lectures. Springer, Berlin, pp 63–71
95. Isanaka BR, Mukhopadhyay T, Varma, RK, Kushvaha V (2022) On exploiting machine learning for failure pattern driven strength enhancement of honeycomb lattices. Acta Mater 239:118226. https://doi.org/10.1016/j.actamat.2022.118226
96. Liu YJ, Chen XL (2003) Evaluations of the effective material properties of carbon nanotube-based composites using a nanoscale representative volume element. Mech Mater 35:69–81. https://doi.org/10.1016/S0167-6636(02)00200-4
97. Wertz J, Homa L, Welter J et al (2018) Gaussian process regression of chirplet decomposed ultrasonic B-scans of a simulated design case. AIP Conf Proc 1949:130007. https://doi.org/10.1063/1.5031602
98. Wang Y, Zhang Y, Zhao H et al (2018) Identifying interphase properties in polymer nanocomposites using adaptive optimization. Compos Sci Technol 162:146–155. https://doi.org/10.1016/j.compscitech.2018.04.017
99. Ahmad I, Basheri M, Iqbal MJ, Rahim A (2018) Performance comparison of support vector machine, random forest, and extreme learning machine for intrusion detection. IEEE Access 6:33789–33795. https://doi.org/10.1109/ACCESS.2018.2841987
100. Das S, Chattopadhyay A, Srivastava AN (2010) Classifying induced damage in composite plates using one-class support vector machines. AIAA J 48:705–718. https://doi.org/10.2514/1.37282

101. Tien Bui D, Tuan TA, Klempe H et al (2016) Spatial prediction models for shallow landslide hazards: a comparative assessment of the efficacy of support vector machines, artificial neural networks, kernel logistic regression, and logistic model tree. Landslides 13:361–378. https://doi.org/10.1007/s10346-015-0557-6
102. Datta A, Augustin MJ, Gupta N et al (2019) Impact localization and severity estimation on composite structure using fiber Bragg grating sensors by least square support vector regression. IEEE Sens J 19:4463–4470. https://doi.org/10.1109/JSEN.2019.2901453
103. Khan A, Kim N, Shin JK et al (2019) Damage assessment of smart composite structures via machine learning: a review. JMST Adv 1:107–124. https://doi.org/10.1007/s42791-019-0012-2
104. Zhang J, Wang Y (2020) Evaluating the bond strength of FRP-to-concrete composite joints using metaheuristic-optimized least-squares support vector regression. Neural Comput Appl. https://doi.org/10.1007/s00521-020-05191-0
105. Yang Z, Gu XS, Liang XY, Ling LC (2010) Genetic algorithm-least squares support vector regression based predicting and optimizing model on carbon fiber composite integrated conductivity. Mater Des 31:1042–1049. https://doi.org/10.1016/j.matdes.2009.09.057
106. Jac Fredo AR, Abilash RS, Femi R et al (2019) Classification of damages in composite images using Zernike moments and support vector machines. Compos B Eng 168:77–86. https://doi.org/10.1016/j.compositesb.2018.12.064
107. Mardanshahi A, Nasir V, Kazemirad S, Shokrieh MM (2020) Detection and classification of matrix cracking in laminated composites using guided wave propagation and artificial neural networks. Compos Struct 246:112403. https://doi.org/10.1016/j.compstruct.2020.112403
108. Meng L, Zhang J (2020) Process design of laser powder bed fusion of stainless steel using a Gaussian process-based machine learning model. JOM 72:420–428. https://doi.org/10.1007/s11837-019-03792-2

Image Processing and Machine Learning Methods Applied to Additive Manufactured Composites for Defect Detection and Toolpath Reconstruction

Guan Lin Chen and Nikhil Gupta

1 Introduction

Additive manufacturing (AM) plays a crucial role in the fields such as aeronautical, automotive, and medical [1, 2], providing possibilities for low cost parts, highly customizable designs and small production runs [3, 4]. AM relies on a variety of software tools and cloud resources to make it a cyber-physical system [5, 6]. For AM, a wide spectrum of feed materials is available across the entire material usage of polymers, metals, ceramics and composites [7–9]. Parts made of glass and carbon fiber filled polymer matrix composites (PMCs) are now being 3D printed as [10, 11]. It is reported that the parts printed by different tool paths can have different properties because of directionality in material orientation and defects [12]. In case of composite materials, the 3D printing toolpath can be used as a method to orient the fibers in a certain direction and obtain a customized part [13]. AM technologies have improved significantly in recent years; however, there still exist numerous challenges in obtaining high quality, resolution and surface finish required for many applications [14–16]. For example, in powder bed fusion, uniformity in the packing of bed from one layer to the other is important for optimizing the processing parameters, which controls the porosity of the powder bed so that the final part has uniform and maximum density [9]. During fabrication of a complex shaped object using material extrusion methods, which is one of the most commonly used by AM methods [17], an outline is printed first to more precisely define the shapes and then an infill pattern is used to deposit material within the contour's shape [18]. This kind of space filling mechanism leads to gaps at the end of the deposited lines and causes porosity if the process is not well optimized, which affects the mechanical properties of the printed part [19]. Additionally, CAD models are widely used in the AM to demonstrate a

G. L. Chen · N. Gupta (✉)
Composite Materials and Mechanics Laboratory, Mechanical and Aerospace Engineering Department, Tandon School of Engineering, New York University, Brooklyn, NY 11201, USA
e-mail: ngupta@nyu.edu

V. Kushvaha et al. (eds.), *Machine Learning Applied to Composite Materials*,
Composites Science and Technology, https://doi.org/10.1007/978-981-19-6278-3_2

digital model and also used for G-code transformation for 3D printers to manufacture the part [20]. However, the curvatures present in the CAD models are often reformed with linear segments in file formats such as STL and can result in the loss of dimensional accuracy in the printed part [5, 21]. These effects cannot be avoided when using material extrusion methods, leading to porosity and micron-sized voids near curvatures, especially in parts of complex geometry and curvatures. The imperfections can be visualized using both destructive and non-destructive imaging methods.

A variety of methods, such as manual or computer assisted non-destructive inspection methods are used to perform quality assessment, which is quiet challenging in the AM domain [22], whether during the process or post-processing to maintain high quality as well as the high manufacturing throughput [23, 24]. In AM, parts are printed in several hundred layers, which can not only be used for locating defects but also can be used to hide specific information within layers [25–27]. The layup process of composite manufacturing is somewhat similar to the layer by layer manufacturing of AM and the machine learning (ML) trained model can be applied to composite quality assessment [13, 28]. Artificial Neural Network, which is inspired by biological neural networks, showing efficiency in dealing with intricate nonlinear behavior and has a strong physical foundation for use in the materials science field [29]. ANN has been used in prediction and optimization of material properties [30], especially in the AM domain, which has a strong process-structure–property relationship [31]. ML methods are now being applied to a variety of problems in materials science, including fields such as development of battery materials and testing of composite materials [32, 33] where enormous amounts of raw data are generated on a daily basis that can be used to train the ML models for effective decision-making [34]. ANN is a computational ML method inspired by the architecture of biological neural networks [35] that has been applied in many problems related to composite materials [36–39] and other materials for defect detection [40, 41].

Micro-CT (μCT) scan, which has been used in the medical field for decades, provides a large image database and is finding increasing applications in the characterization of AM parts. Such large image sets are a limitation for manual inspection methods, but an asset for training of ML methods [42–45]. It is beneficial to pre-process the meaningful features that are used for prediction to reduce the training effort and increase the accuracy of the designed ML algorithm. The location and orientation of the reinforcing fibers govern the mechanical response of composite materials, which can be detected by ML methods [46]. ML methods have also been used on optical images to identify and classify two-dimensional materials [47]. The use of ML methods on large image databases sometimes requires significant signal processing effort for making the search faster for implementation in the real time defect detection systems.

In this work, describes in detail the steps required for image processing, ANN model training and validation, and defect detection using a database of images obtained on a 3D printed composite material specimen. A glass fiber reinforced PMC filament is used to fabricate specimens by a commercial material extrusion 3D printer. The specimens are imaged using a μCT scanner. The image dataset is used to train a ML algorithm. Previous studies revealed that 2D images with irregular outlines increase the difficulty in analyzing the microstructure features of composite materials such as the fiber orientation identification [13, 47], such limitations are overcome by cropping the training images in circular shape during training. Application of image processing methods to reduce the size of the dataset helps in making the training significantly faster. In addition, two kinds of ANNs, namely Convolutional Neural Network (CNN) and Recurrent Neural Network (RNN), with three different architectures—one-dimensional and two-dimensional algorithms in CNN, and Long-Short-Term-Memory cell in RNN and its related Python codes are presented, and the implementations of the three trained model and the resulting tests are presented in this work.

2 Methods

2.1 Sample Preparation

The samples are designed using SolidWorks 2017 (Waltham, MA) and saved in STL format. The shape of the sample is a cube with 6 mm side. ABS-GF10 glass fiber reinforced acrylonitrile butadiene styrene (ABS) filament of 1.75 mm diameter, manufactured by 3DXTECH, Grand Rapids Michigan, USA, is used for 3D printing. ReplicatorG software is used for preparing the sliced model and generating the G-code. The printing parameters included 100% solid infill, feed rate of 41 mm/s, extrusion temperature of 220 °C, and build platform temperature of 120 °C. A FlashForge-Creator Pro Dual Extruder printer is used for printing the specimens.

2.2 μCT Scan Image

A Skyscan 1127 (Bruker, Belgium) μCT scan system is used at source voltage of 44 kV, current of 222 μA and the camera pixel size of 9.5 μm with a rotation step of 0.6° per scan, for 360° rotation. Normal scanning time is about 40 min with lowest resolution. However, larger specimens can take 6–8 h in one scan based on the image resolution. Each scan generates thousands of images, which are usually saved in tiff format.

The images acquired from μCT scan were reconstructed using NRecon software. In this step, the image can be tuned with function such as "smooth", which can improve the clarity of the image slice, "contrast", which can reduce the noise and "rotation", which is used to rotate at a fixed angle to output the reconstructed images with desired angle as Fig. 1 shown. Since, the sample was randomly mounted on the specimen plate and the scanning process involved rotation, the printing direction of the output images was random with respect to the scanning direction. It is important to designate a reference angle so that all the images can be interpreted with respect

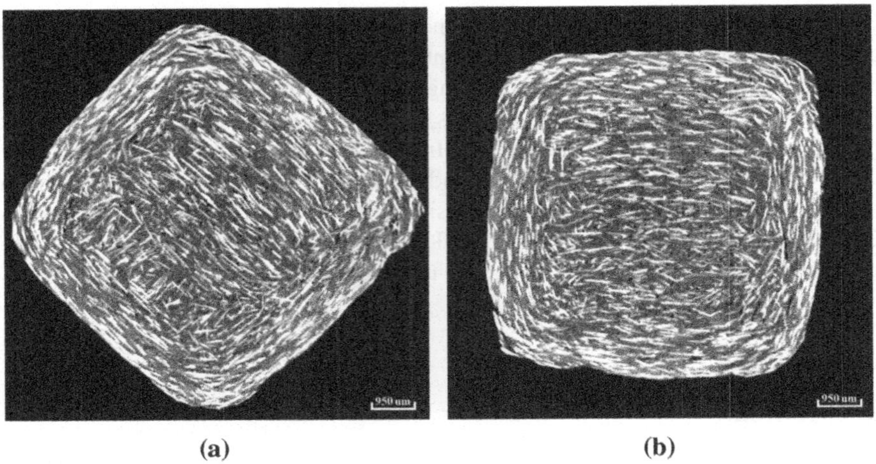

Fig. 1 An example of a μCT scan output image with **a** random angle and **b** after rotation to 0°

to the same reference framework. Since fibers are included in the feed material and the fibers provide high contrast, they are used for identification of the direction. In this work, 0° is defined as horizontal reference. Therefore, the reconstructed images are rotated using the "rotation" function in NRecon software to output the images to be the reference images of 0°. Once the images are rotated, they can be directly compared. The present work is exploratory for developing the image processing and ML model training methods; hence, the geometry of the 3D printed part is kept simple.

The square shape of the specimen having four sharp corners is a consideration in this work. The sharp corners would inevitably affect the accuracy for ML because any sharp angles in the image may cause bias while training the model, resulting in inaccurate predictions. To increase the accuracy of the trained model, the data selection and features extraction are the key steps. Therefore, it is better to sub-segment each μCT scan image slice into smaller images and use circular shape sub-images. Rotation of circular sub-images will preserve the features in the circular area without any interference at the edges. The model trained with a circular image dataset is still capable of predicting a square image. However, if the model is trained on a square image dataset, it will not be able to predict a circle image with accuracy.

2.3 Circular Image Dataset Preparation for ML Model Training

Two prediction purposes are introduced. One is to build a ML model capable of predicting the printing direction and its angular information in each layer of the sample, which can then provide the information of the printing angles used in the 3D printer. Another purpose is to predict the movement of the print head in each layer, which can then be used to analyze the printing signature throughout the printed sample.

There are two different methods to obtain the required dataset. The first method uses image reconstruction after the μCT Scan and the second method uses a Matlab code. Both of the methods will be discussed thoroughly in the following section.

2.4 Dataset for Orientation Prediction in Each Layer

In this section, the prediction of the overall printing orientation in each layer of the sample is developed. Circular sub-images are first extracted from each image slice as shown in Fig. 2. The function of "region of interest" is used during the image reconstruction with NRecon software to capture a certain area of the scanned image. Figure 2a shows an image where glass fibers seem to be oriented in the horizontal direction. Figure 2b, c show two sub-images that contain fibers oriented in multiple directions. These features appear from two different slices because the scan resolution is much finer than the deposited layer thickness and the specimen shows some warpage due to thermal contraction upon cooling. Such intermixing of features may also take place due to mismatch between the printing plane and CT scanning plane.

Depending on the resolution setup for the μCT scan and the specimen size, the total numbers of images will be different in each dataset. For the specific sample used in this work, there are 663 images obtained after the μCT scan.

Next, a reference angle needs to be decided. In this work, images with horizontal glass fiber are designated as 0°. For training the ML algorithm, a sufficiently large database is necessary. Moreover, to distinguish the fiber angle, the model needs to be able to predict fibers along any angle in the entire 360°. However, it actually only needs to predict 180° to indicate the whole 360°, since 10° can represent 190° and

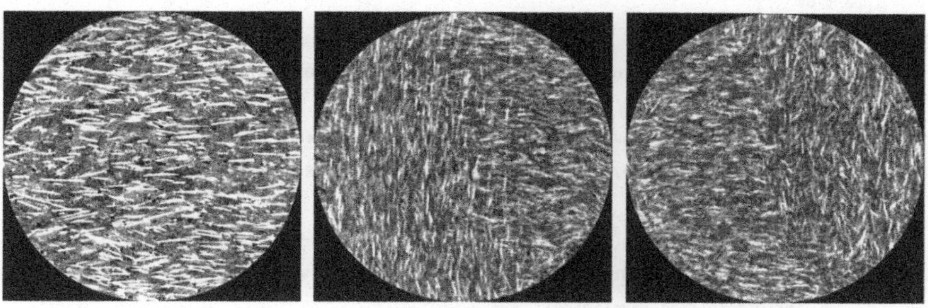

(a) (b) (c)

Fig. 2 Sub-images extracted from a μCT scan image slice

0° is the same as 180°. Therefore, in order to train the model, images with clear visibility of 0° fiber angle are synthetically rotated counterclockwise at intervals of 1° each from 0° to 180° using a Python code shown as Fig. 3 to get the images database to train the model for predicting various rotation angles. The rotation of the images can be conducted at smaller steps but the discontinuous short fibers in the present case are not oriented exactly in the same direction. Some local variations exist within one extruded line as well as some fibers are bent. Such conditions will provide false positives. Hence, the rotation step is kept at a higher angle. The process allows creating a dataset with controlled orientation of the glass fiber to train a model for printing orientation prediction in each layer.

A few μCT images showing overlapping printing directions in Fig. 2b, c are mostly at the boundary of two deposited layers due to the slight slope that can be either in the printing or in the specimen positioning on the μCT stage. These images are removed from the dataset in the present case because there is enough information available from other slices that are cleaner. In order to have a defined/labeled angle, a set of 207 images that show clear orientation of fibers is identified and rotated so that the fiber orientation is horizontal. These images are labeled as 0° fiber orientations. All the images are cropped with a region of interest to decrease the number of pixels and remove bias caused by the sharp angles in four corners for machine learning training. The cropped images, shown in Fig. 2, have a pixel size of 536 × 536. Individual fibers in each image show some variation in their direction. However, the algorithm takes the global signature as the features for the 0° and disregards the individual fiber orientation. Each of these images are then rotated counterclockwise from 0° to 180° at 1° interval using a Python code. This procedure allows creating a large training database with controlled fiber orientation and trains the model to identify any angle. This procedure resulted in 37,467 images. The process of synthetically rotating the images and cropping them with built-in function during image reconstruction with

```python
# Rotate the Images and save to different dir
from PIL import Image
import matplotlib.pyplot as plt
import os
import shutil
path = 'C:/Users/User/Desktop/Guanlin_CNN1D/CNN1D/2D book 207 testing' # The path of 0 deg reference images
for deg in range(181): # Save rotated image into folder named 0, 1, 2, 3,...,180
    out_put ='C:/Users/User/Desktop/Guanlin_CNN1D/CNN1D/0_180 book testng/'+str(deg)
    print(out_put)
    for image in os.listdir(path):
        img = Image.open(path+'/'+image)
        img.rotate(deg).save(image)
```

Fig. 3 Python code for rotating image form 0° to 180° at 1° interval and save to folders named as 1, 2, 3…180

NRecon software used to generate training data is known as image augmentation. This process is helpful in training the ANN model to be robust.

2.4.1 Dataset for Toolpath Prediction in Each Layer

In this section, training dataset preparation is slightly different compared to the previous section. Here, the purpose is to determine the printing path, or toolpath, of each layer of the sample with specific image features for the model training. Moreover, the algorithm used for this dataset is 1D and RNN, which needs to transfer the features contained in a 2D image into a meaningful dataset with 1 dimensional format. Therefore, the training dataset should contain meaningful features, which represent the movements of the print head. Thus, one layer of image with clear fiber orientation is used to conduct the feature extraction for model training dataset preparation. Here, instead of using region of interest to extract the features from a certain area, for toolpath prediction, the whole image slice is preferable since the toolpath of the whole layer and the whole sample need to be predicted to completely recreate the toolpath of the sample.

Each μCT image slice represents a variety of fiber directions based on the space filling algorithm used for determining the print head movement. The ML model tuning dataset needs to identify these features, Hence, each image is sliced in circular sub-images of 100 × 100 pixels. The sub-sampling process and the resulting images are shown as Fig. 4. This process results in 961 cropped circular images for each μCT image slice. The circular shape helps in reducing the error caused by irregular outline of the specimen. In this work, 150 images with clear visibility of 0° fiber angle are selected and then the images are synthetically rotated counterclockwise at intervals of 1° each from 0 to 180° using a Python code shown as Fig. 3. Similar to the previous section, this cropped circular images for each μCT image slice help in reducing the error caused by irregular outline of the specimen. Then, 5 images are randomly selected to be the test image set, 115 images are randomly chosen to be the training data and the rest 30 images are used to be the validation images. Three of them are then rotated with Python code to create the training, validation, and test dataset with total images of 27,150 images for model training and testing. A special image lossless-processing algorithm called BSIF then used to convert the 2D images into a 1D numerical dataset for reducing the computing power and increasing the processing speed. The detail of the BSIF process are discussed in the next section.

2.4.2 Binarized Statistical Image Features (BSIF) Algorithm

Binarized Statistical Image Features (BSIF) algorithm is used to convert an image into a binary image format without losing valuable features [48]. Figure 5 shows an example of the image produced by the BSIF algorithm. The images processed through BSIF algorithm are used for training, validation and testing the ML algorithm. The output image is a binary code for each pixel in the image and is stored in a format

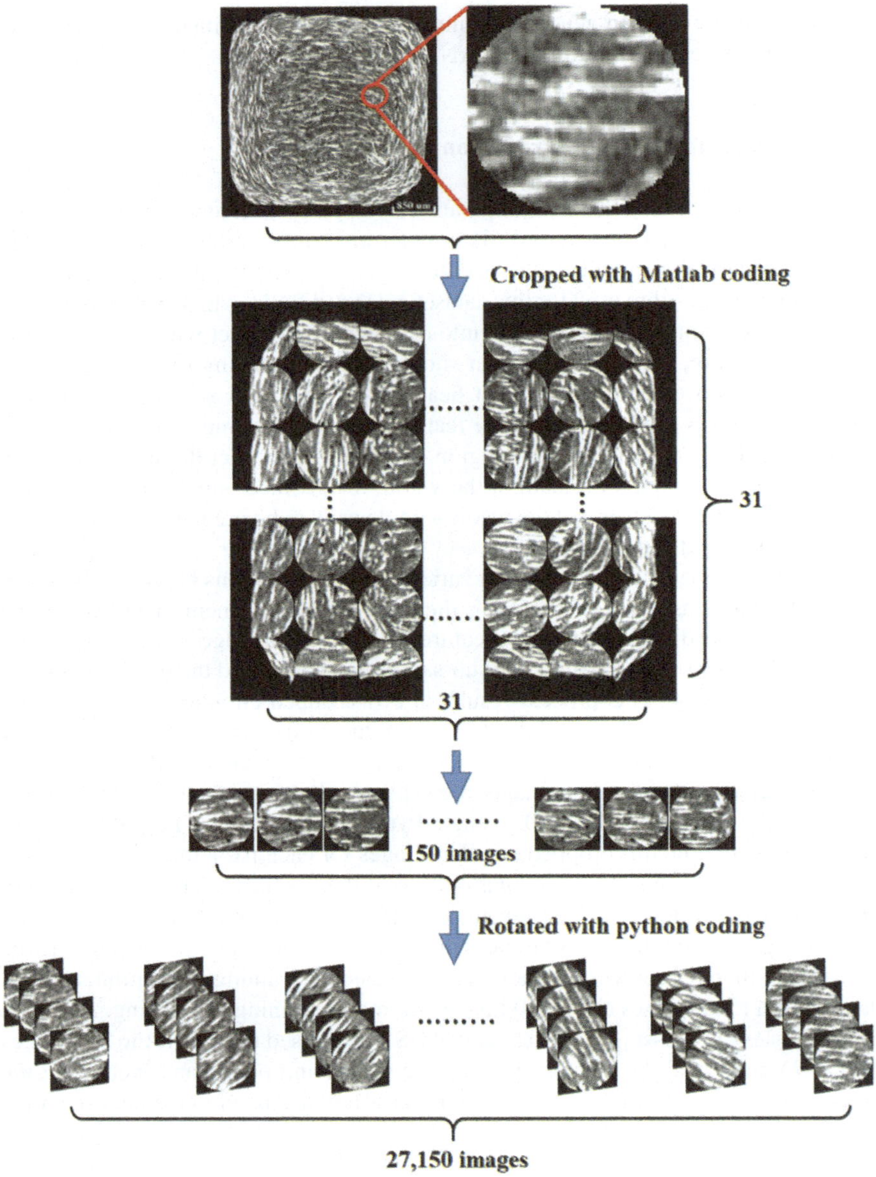

Fig. 4 The image dataset (27,150 segmented images) obtained from μCT scan after removing overlapping images and being labeled as 0° to 180° for ML

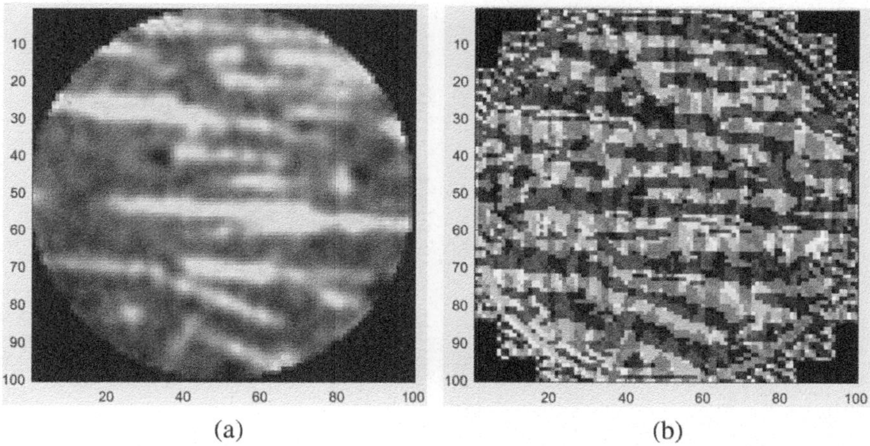

Fig. 5 BSIF representation of the glass fiber orientation image with 100 × 100 pixels. **a** The original circular cropped 2D image and **b** the 2D image converted by BSIF

of 1D numerical array, which makes it convenient to handle large amounts of data. Although the visual representation of features in Fig. 5 is not well resolved, such images are proven to perform well in ML. Use of BSIF code reduced each image to 1 × 256 numerical data as shown in Table 1, which exponentially reduces the computational expense involved in running the ML algorithm. Since 27,150 images are used, the resulting data was saved as a csv file with 27,150 rows of features, which is extremely helpful for ML. This procedure uses a large dataset but the training time for the ANN is significantly reduced without losing the accuracy for interpreting the features present in the image [48]. The 2D images are converted into 1D numerical data with one row and 256 columns and since the more meaningful features represent an output the higher accuracy of the model can be obtained.

In this work, only the dataset for toolpath prediction in each layer (introduced in Sect. 2.4.1) needs to be converted with BSIF since the 2D CNN needs original 2D images as its input data.

2.5 *Machine Learning*

Two kinds of ANN architectures are used to build models using Python with Tensorflow: RNN and CNN. Both neural networks can deal with sequences with variable lengths. The RNN uses memory-state to process the input data on each neuron. In this work, an RNN architecture with 5 layers and 64 Long Short-Term Memory (LSTM) cells is used to train the model. A CNN model is trained with total 5 layers and kernel sizes 5 and filter size 80 in the first layer and the second layer to iterate through the data to train the model.

Table 1 Image data converted into 1 row of numerical data after BSIF process

0	1	2	3	4	…	251	252	253	254	255
0.1832	0.003904	0.001528	0.00112	0.004256	…	0.00428	0.001108	0.001192	0.001904	0.00552

1 Row × 256 Columns

The original images are used for 2D CNN algorithm because it has its own image conversion process built in the algorithm. On the other hand, for 1D CNN and RNN, the images converted with BSIF are loaded and rearranged in the form of an array as (20,815, 1, 256) for training feature. This implies that this training set has 20,815 data representing the true angle label, with time-step as 1, and 256 features points. The same process applies to validation dataset, which results in an array of (5430, 1, 256) and the test dataset has the numpy shape of (905, 1, 256) for RNN algorithm and (20,815, 256, 1), (5430, 256, 1), (905, 256, 1) for training, validation and testing, respectively, in 1D CNN algorithm. The loss function depends on the desired result parameter such as the fiber angle. Hence, the mean square error (MSE) was used in both RNN and 1D CNN as the loss function in the model. The predicted values and the actual value of the angle are used to calculate MSE.

In summary, this work uses three different ML algorithms. 1D and 2D CNN and RNN are compared for their accuracy. Each of them target different dataset and loss function, the computer language used is python platform with Tensorflow in Windows 10 environment.

2.6 The Architectures of the Machine Learning Algorithms

2.6.1 2-Dimensional Convolutional Neural Network (2D CNN)

The 2D CNN ML algorithm uses Python platform, with Tensorflow, CNN (Convolutional Neural Network). CNN works well for identifying simple patterns within the data, which are then used to form more complex patterns within higher layers. CNN is very effective in deriving interesting features from shorter (fixed-length) segments of the overall data set and where the location of the feature within the segment is not of high relevance.

In 2D CNN algorithm, the input is the original image data and the ML algorithm convert the image features in its own system. The loss function in 2D CNN is "sparse categorical cross entropy", which is used to distinguish different categories. In 2D CNN, the model can be used to predict the printing orientation in each layer as a classifier. Hence the predicted values represent the orientation of each layer, and can iterated through whole sample to provide the information of the printed direction. However, it is not capable of identifying the toolpath in each layer. Also, since it uses the original images as the input, the processing needs a huge computing power. Additionally, when increasing the number of categories, the training time increases substantially. Since the process uses original images and the in-built image processor provides a lossless conversion procedure, the accuracy can reach almost 99% with proper setup parameters. Thus, the 2D CNN is good at image analysis, especially if high performance computing facilities are required. In this section, a 2D CNN model and the process steps will be presented in detail.

The dataset for orientation prediction in each layer (Sect. 2.4), which contains 207 clear views of 0° orientation images is used to create the input data. The images need to be saved in folders according to the angle they represent, which means each classification category is saved in corresponding folder. The 0° reference images are split into 3 subsets of images, one for training, one for validation and one for testing. Here, 7 images are chosen to be the test data, and the ratio of training and validation data is set as 8:2. Thus, the rest of the 200 images are split into 166 images for training and 34 images for validation. With the help of python code, as Fig. 3 shows, the split images are rotated from 0° to 180° with 0° interval and saved to folders named as 1, 2, 3…180 for training, and the same process is conducted to create validation dataset and test dataset. By doing so, the training dataset with 30,046 images and the validation dataset with 6154 images and the test dataset with 1267 images are obtained and each angle of images is saved to the corresponding folder.

Once the training, validation and test dataset are prepared, they are fed into the 2D CNN algorithm for model building. The image dataset needs to be re-sized from 536 × 536 to 100 × 100 pixel to reduce the needed computing power and then create the training and validation datasets. Last, the training and validation datasets need to be appended to corresponding features and labels, which are then converted into numpy arrays with the shape of (30,046, 100, 100, 1) for training dataset, (6154, 100, 100, 1) for validation dataset in order to fit the 2D CNN model training algorithm. The first number represents the number of images, the second and the third numbers represent the pixel size in x and y directions, and the fourth number represents the number of images in each sequence. The Python code used for the 2D CNN image processing is shown as Fig. 6.

The architecture of the 2D CNN algorithm is shown as Fig. 7, which has 5 layers. The filter size used is 80, and the kernel size in layer 1 is (5, 5) and (3, 3) in layer 2. The Maxpooling size is (2, 2) in layer 1 and 2. The hidden layer Dense is 200. The output layer has 181 categories, so output Dense is set as 181. The activation is selected as "softmax". Batch size iterated is 100, and the epochs is set as 50. Also, a Dropout function of 0.1 is used to intentionally drop 10% of the training data in each training epoch to prevent the overfitting.

The setup parameters used in this 2D CNN algorithm are obtained by trial-and-error and the parameters are extremely data-dependent. Therefore, different numbers of image, different pixel size of image, and even different shapes of the image used for training will need to find the suitable parameters accordingly. A checkpoint function is used to monitor the accuracy during each epoch, and save the best model throughout the whole training process. The total time for the training is 24 h.

3 2D CNN Result

When the training process is completed, first thing is to check the fitting status during the training process, whether it is overfitting or underfitting. To do that, a recall function in Python is used and the training and validation histories are plotted as Fig. 8. The result in this training process shows no overfitting or underfitting, since the validation result and the training result have the same trend and do not show large

```
# Resize the image to speed up the process
Img_Size = 100
# Training_dataset preparation
training_data = []
def create_training_data():
    for category in CATEGORIES:
        path = os.path.join(DATADIR, category)
        class_num = CATEGORIES.index(category)
        for img in os.listdir(path):
            img_array = cv2.imread(os.path.join(path,img), cv2.IMREAD_GRAYSCALE)
            new_array = cv2.resize(img_array, (Img_Size, Img_Size))
            training_data.append([new_array, class_num])
create_training_data()
# Validating_dataset preparation
validating_data = []
def create_validating_data():
    for category in CATEGORIES:
        path = os.path.join(DATA_valid, category)
        class_num = CATEGORIES.index(category)
        for img in os.listdir(path):
            img_array = cv2.imread(os.path.join(path,img), cv2.IMREAD_GRAYSCALE)
            new_array = cv2.resize(img_array, (Img_Size, Img_Size))
            validating_data.append([new_array, class_num])
create_validating_data()
# Re-shape the Training dataset
X=[]
y = []
for features, label in training_data:
    X.append(features)
    y.append(label)
X = np.array(X).reshape(-1, Img_Size, Img_Size, 1)
y = np.array(y).reshape(-1, Img_Size, Img_Size, 1)
print('X shape :'+str(X.shape))
# Re-shape the Validating dataset
X_valid = []
y_valid = []
for features, label in validating_data:
    X_valid.append(features)
    y_valid.append(label)
X_valid = np.array(X_valid).reshape(-1, Img_Size, Img_Size, 1)
y_valid = np.array(X_valid).reshape(-1, Img_Size, Img_Size, 1)
```

Fig. 6 The python code used for resizing images and dataset preparation in 2D CNN algorithm

```
# 2D Convoultional Neural Network architecture
X =X/255
X_valid = X_valid/255
model = Sequential()
# Layer 1
model.add(Conv2D(80,(5,5), input_shape = X.shape[1:] , padding ='SAME'))
model.add(Activation("relu"))
model.add(MaxPooling2D(pool_size=(2,2)))
# Layer 2
model.add(Conv2D(80,(3,3), padding ='SAME'))
model.add(Activation("relu"))
model.add(MaxPooling2D(pool_size=(2,2)))
# Layer 3
model.add(Dropout(0.1))
model.add(Flatten())
# Hidden Layer
model.add(Dense(200))
model.add(Activation('relu'))
# Output Layer
model.add(Dense(181, activation='softmax'))
model.compile(loss='sparse_categorical_crossentropy',optimizer="adam",metrics=['accuracy']
)
filepath = 'C:/Users/User/Desktop/2D CNN model-{epoch:02d}-{val_acc:.3f}.model'
checkpoint = ModelCheckpoint(filepath, monitor='val_acc', verbose=2, save_best_only=True,
mode='auto')
model.summary()
history = model.fit(X, y, batch_size = 100, epochs=50, verbose=1,
                    validation_data = (X_valid, y_valid), callbacks=[checkpoint])
```

Fig. 7 2D CNN architecture for the classification model

variation. If the graph shows a large gap between the two lines, it indicates that the model is not trained properly and the setup parameters need to be further tuned. The acceptability of the result also depends on the desired accuracy. Typically, in 2D CNN machine learning, the accuracy of training is often higher than the accuracy of validation, which is also observed from the resulting plot shown in Fig. 8.

To test the accuracy of the trained model, the test dataset is needed, which has been separately prepared with 1267 data using the same preparation process as the training dataset. An array shape of (1267, 100, 100, 1) is required to be compatible with the training dataset. In this 2D CNN training, the shape used for training is (30,046, 100, 100, 1). Therefore, the test dataset must have the same pixel size of 100×100. Then the folders with images are loaded in the Python code, and the output categories, which contains 0° to 180°, are created. A function "model.evaluate" is used to check

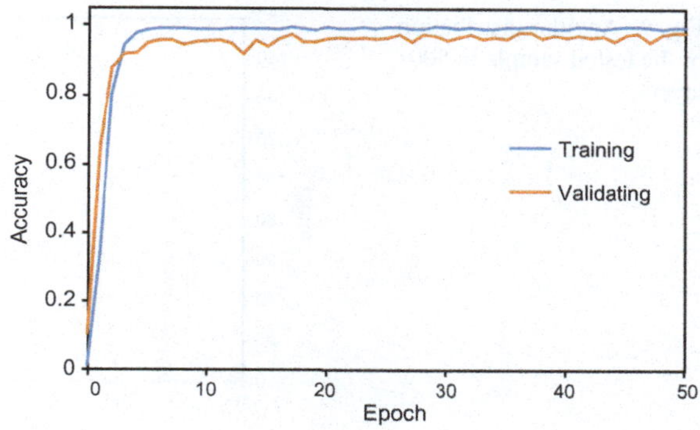

Fig. 8 Training and validation history in each epoch for 2D CNN model

the model accuracy. By doing so, the trained model with the accuracy of 0.9684 is achieved, which can be used in a practical situation.

Next, the prediction accuracy of the trained model is determined by using a set of images of unknown angular information. Here, another 3D printed specimen of a similar type is imaged by μCT Scan and 500 images with unknown angular information are obtained. The images are then renamed according to its layer height from the bottom of the sample to its top with the name of "Layer 1", "Layer 2", etc. These 500 images are then saved in a folder and then loaded into Python code. Next, the images are used to create a test dataset and converted into a numpy array with the shape of (500, 100, 100, 1). The names of the images are appended to the "Layer" column, and the converted images' features are used for prediction. The results are appended to the "Predicted Angle" column. Last, the outcome is saved in the tabular form and the predicted angle distribution for each layer is plotted, as shown as Fig. 9, which can reveal the printing orientation of the tested sample for each layer. Here, the predicted angles are located around 90° and 0° showing that the sample is printed with these two angles repetitively. One major benefit is that the prediction process is very fast. The dataset has 500 images and the time taken for prediction is less than a second. Therefore, even though it takes 24 h to train the model, the time required for prediction is extremely short.

3.1 1-Dimensional Convolutional Neural Network (1D CNN)

In this section, a 1-D ML algorithm is introduced. The 1D CNN, which means the input data is 1-Dimensional. Therefore, 2D images must go through the BSIF algorithm to be converted into 1D data. In the 1D CNN algorithm, a loss function, Mean Square Error (MSE) is used. Each image features are converted with the BSIF lossless-process into a numerical data with 1 row and 256 columns, then in order to easily extract the needed features and to designate the angle/label, a column, named "Image Angle" is added as the first column to designate the angle/label and

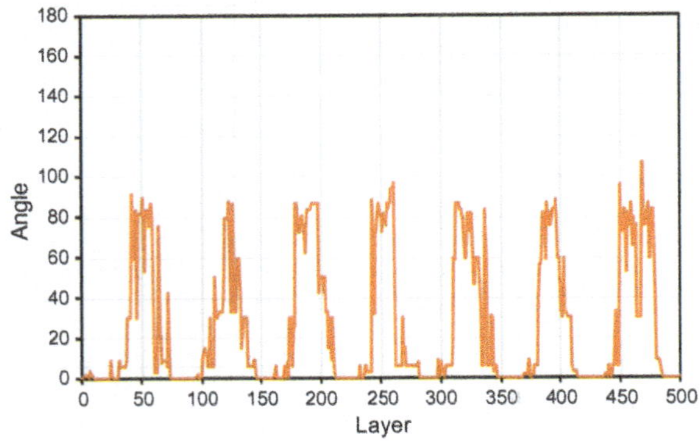

Fig. 9 Angular distribution of the tested sample in 500 layers

"Feature_1", "Feature_2", all the way to "Feature_256" are add at the top of each feature. Thus, all the information representing each image in a fixed-length and its angle/label and features are saved as a CSV file.

The 1D CNN algorithm is capable of predicting the printing orientation in each layer. It is also capable of predicting the toolpath of the sample in each layer. The database containing 27,150 images needs to be divided into training, validation, and test datasets. Since 150 images with clear view of 0° fiber orientation are selected as described previously, the images are used to rotate 1° interval counterclockwise from 0° to 180° to represent 180 angles. Moreover, in order to avoid bias during the model training, 115 images are picked to use as the training dataset for each angle, which total to be 20,815 sub-images. Then 20 images from each angle are added to the validation dataset, which sums up to 5430 sub-images. Finally, 905 sub-images are used in the test dataset. By feeding the dataset with evenly distributed weights, the ANN will have the least bias when the training is completed. In 1D CNN, the loss function is "Mean Square Error", and the output is a value representing the predicted angle. The CSV files for training data, validation data and test data are loaded using the "panda" function in Python to train the model.

The dataset needs to be prepared to form a correct format after being loaded in order to feed into CNN algorithm. The Python code used is shown as Fig. 10, where "data.values" function is used to capture the features within the data. The first colon in the bracket means the function iterates through every row's data, and the number after the comma represents the data in each number of the column. Here, "2:" means obtaining the data from the second column all the way to the last column and then the value is used as training features. Then, the output labels need to be defined. "[:, 1]" is used to obtain the data in the first column for every row and the value is defined as training labels. The same process is used for the validation dataset to obtain the validation features and labels.

After the process of iterating through every row and column of the CSV file, a re-shape function is used, which creates a dataset with a shape of (20,815, 256, 1) for training feature dataset, (5430, 256, 1) for validation feature dataset and (905, 256, 1) for testing feature dataset. Here, the first number represents the number of the

```
# Training and Validating Features and Labels Designation
X_train = data.values[:, 2:]
Y_train = data.values[:, 1]
X_valid = val_data.values[:, 2:]
Y_valid = val_data.values[:, 1]
# Reshape the nparray to fit the training and testing model
X_train = np.array(X_train, dtype=np.float).reshape(X_train.shape[0],X_train.shape[1],1)
Y_train= np.array(Y_train, dtype=np.float).reshape(Y_train.shape[0],1)
X_valid = np.array(X_valid, dtype=np.float).reshape(X_valid.shape[0],X_valid.shape[1],1)
Y_valid= np.array(Y_valid, dtype=np.float).reshape(Y_valid.shape[0],1)
```

Fig. 10 Python code used to prepare the training and validation dataset in 1D CNN

data/images, the second number represents the features each data has, and the third number represents the number of images for each process. As for training, validation and test label datasets, the shape of the array is (20,815, 1), (5430, 1) and (905, 1) respectively. The first number represents the number of label data, and the second number represents the number of images for each process.

For the output of an angle prediction, it would be easier to understand when the output is a value that represents the predicted angle. Hence, the loss function "Mean Square Error" (MSE) is used. The model can output a number value to be the representation of an angle for the input feature.

The architecture of the 1D CNN is shown as Fig. 11. Similar to 2D CNN architecture, the checkpoint function is used to capture the model with the lowest MSE value throughout the training process. There are 5 layers used in the 1D CNN architecture. The filter size for layer 1 and 2 is 256, which means the whole 256 features are considered, the kernel size is 1, which means 1 full image features are processed at each time. Padding used is "same", the activation used in each layer is "relu", the hidden layer has Dense of 200 and the loss function is "Mean Square Error", the optimizer is Adam, which provides a learning rate of 0.001, the batch size is set as 128, and total 10,000 epochs is used. The verbose in checkpoint is set as 2, which means the process will just mention the number of epochs, and in "model_m.fit" the verbose is set as 1, which will show an animated progress bar for the user to observe if the model is properly trained. The parameter setup in this training is also extremely data-dependent, therefore, with different datasets, the parameters need to be tuned again.

3.2 *1D CNN Result*

After completion of the model training, the first thing is to check the training history to make sure it has no overfitting, which shows the validation curve to be much lower than the training curve or underfitting, which shows the training curve to be much

```
# 1D CNN Convolutional neural network architecture
model_m = Sequential()
# Layer 1
model_m.add(Conv1D(256, (1), activation='relu',
             input_shape=(X_valid.shape[1],X_valid.shape[2]), padding = 'same'))
# Layer 2
model_m.add(Conv1D(256, (1), activation='relu', padding = 'same'))
# Layer 3
model_m.add(Flatten())
# Layer 4
model_m.add(Dense(200, activation='relu'))
# Layer 5
model_m.add(Dense(1, activation='relu'))
model_m.summary()
filepath = 'C:/Users/User/Desktop/models/1D CNN book-{epoch:02d}-{val_loss:.3f}.model'
checkpoint = ModelCheckpoint(filepath, monitor='val_loss', verbose=2,
             save_best_only=True, mode='auto')
model_m.compile(loss= 'mean_squared_error' , optimizer= 'adam' , metrics=[ 'mse' ])
history = model_m.fit(X_train, Y_train, batch_size=128, epochs = 1000, verbose=1,
                  validation_data = (X_valid, Y_valid), callbacks=[checkpoint])
```

Fig. 11 The ML architecture with 5 layer in the 1D CNN algorithm

lower than the validation curve. The training and validation histories are plotted as Fig. 12. The graph shows no sign of overfitting or underfitting. Then, the function "model.evaluate" is used to test the accuracy of the model. Since the MSE loss function is used, the output of the evaluation is the MSE of the prediction. A MSE of 0.9651 is obtained showing a promising result for the CNN implementation.

The model trained in this section is used to predict the toolpath of the composite material specimen image set obtained from μCT scan. In this work, the sample used

Fig. 12 The training and validation MSE of each epoch in 1D CNN model

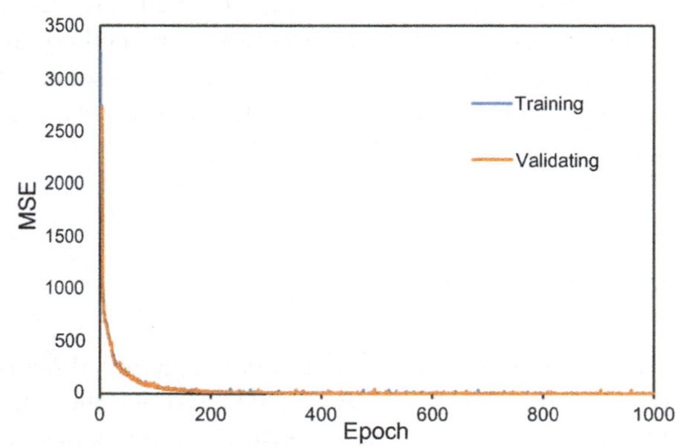

for toolpath prediction is a square cube with 1 cm length on each side. To predict the toolpath of the whole sample, a μCT scan needs to be conducted to obtain the sliced images. The μCT scan image used to perform the toolpath prediction has pixel size of 2657 × 2689. Next, similar to all training dataset preparation, the images are cropped as circles using a Matlab code and after the cropping process, 676 images with pixel size of 100 × 100 are obtained. These 676 images are then converted with BSIF and become 676 data. Each data has 256 numerical features and are saved as a CSV file. Thus, a shape of (676, 256, 1) numpy data array is ready for toolpath prediction.

In order to clearly view the toolpath in each layer, a direction indicator is used. The idea is to impose each small cropped circular image with a direction indicator corresponding to its prediction and then combine all the direction indicators to reconstruct the whole image to represent the layer showing the toolpath. Hence, the predicted result needs to be recorded and saved as a CSV file, which has a column showing the region the cropped image belongs to and a column for prediction result of that region. The direction indicator images showing the angle from 0 to 180° are saved with a file name with respect to its angle. Then, according to the prediction result, a Python code "if and else" is used to match the prediction result and the corresponding image name of the direction indicator. For example, an image representing region 10 is predicted as 44°, then the Python code will match the prediction result 44° to the direction indicator, which is named as 44 and the direction indicator is saved to represent the region 10. After the process iterates through all predictions, a collection of all 676 direction indicator images is saved. Finally, a 26 × 26 grid is created to display all 676 predicted direction indicators are shown as Fig. 13.

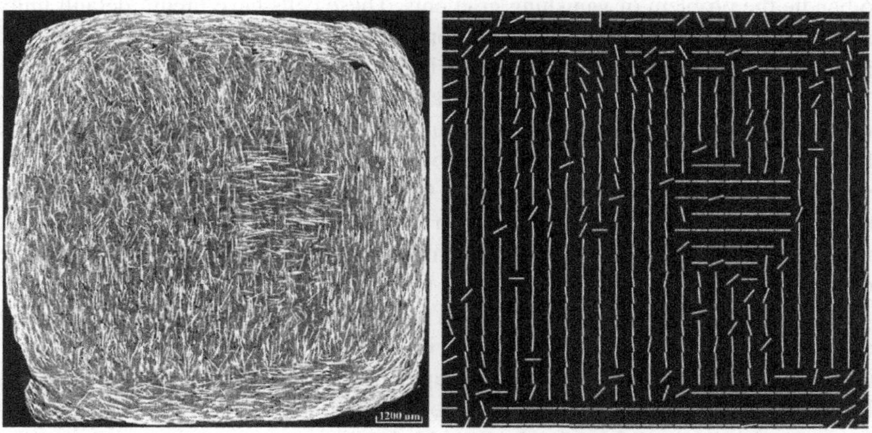

(a) (b)

Fig. 13 Imposed toolpath reconstruction with 1D CNN model

3.2.1 Recurrent Neural Network (RNN)

Recurrent neural network (RNN) is a supervised ML algorithm, which is designed to model sequential data. The order is very important in sequential data. There are different forms of sequence modeling algorithms but the one used here is the many-to-one sequence model, which implies that the input data is a sequence but the output is not a sequence, rather a fixed-size vector. In RNN, the hidden layer has inputs from both the input layer and the hidden layer from the previous step. The flow of information in an RNN from one time-step to another introduces memory of past inputs into the network [49].

The algorithm used here is a multilayer RNN, which is used to predict the direction of fibers in a μCT scan image. The input is the μCT-scan image features obtained after the BSIF process and the output is the fiber orientation angle. At any time instance the model uses the information from the past and input to predict the output. Since AM follows a sequential process of printing, the fiber orientation at each layer can be helpful to predict the orientation of fiber of the next layer. Typically, a backpropagation through time (BPTT) algorithm is used to train an RNN, which sometimes has a problem of vanishing gradient. RNN model faces difficulty in learning the long-term dependencies because it is trained with sequential data, which implies that the model will not be able to relate the images which are captured several time steps apart. To address these issues, the RNN architecture with LSTM network is used [50]. RNN with hidden layers containing LSTM cells takes information from the input and from the previous hidden layers and calculates the output through a set of equations and then sends the information to the next layer in the model and to the hidden layer, namely, another LSTM cell in the next time step. LSTM cells are designed to handle the problem of vanishing gradients. These cells have inbuilt default units programmed to remember the updates from the previous time steps without loss of information over long time steps, making them suitable for large image datasets.

As mentioned previously, the BSIF process has converted features of each image into 1 row and 256 columns. A column named "Image Angle" is added to the file as the first column to designate the angle/label. Thus, all the information to represent each image is available in a single row. Any change to the input data affects the algorithm and the setup parameters need to be tuned. In RNN, the dataset for toolpath prediction in each layer is used, which means a model capable of predicting the toolpath can be acquired. Here, the same training, validation and test data of CSV files as 1D convolutional neural network is used. The Python code used for reshaping is similar as the one used in 1D CNN algorithm, where the only difference is in array arrangement. In RNN architecture, the shape of the training data is (20,815, 1, 256), the validation data is (5430, 1, 256), and the test data is (905, 1, 256). The first number represents the number of dataset the CSV file has, the second number represents the amount of data being iterated in each time step, in this case, 1 image is used in a time step. The architecture of the RNN Python code used is shown in Fig. 14. There are 5 layers in the algorithm. The number of LSTM cells used in first 2 layers is 64, the hidden layer has "Dense" of 128 with only 1 output value, the loss function is "MSE", the optimizer is Adam, which has a learning rate of 0.001 and the batch size

```
# RNN Recurrent Neural Network architecture
model = Sequential()
# Layer 1
model.add(LSTM(64, input_shape=(X_train.shape[1:]), return_sequences=True))
model.add(BatchNormalization())
# Layer 2
model.add(LSTM(64, return_sequences=True))
model.add(BatchNormalization())
# Layer 3
model.add(LSTM(64,))
model.add(BatchNormalization())
# Hidden Layer
model.add(Dense(128, activation='relu'))
# Output layer
model.add(Dense(1))
model.summary()
# learning rate 0.001
model.compile(loss='mse', optimizer='Adam' , metrics=['mse'])
filepath = 'C:/Users/User/Desktop/models/RNN book-{epoch:02d}-{val_loss:.3f}.model'
checkpoint = ModelCheckpoint(filepath, monitor='val_loss', verbose=2,
             save_best_only=True, mode='auto')
history = model.fit(X_train, Y_train, batch_size = 128, epochs=10000, verbose=1,
                    validation_data = (X_valid, Y_valid), callbacks=[checkpoint])
```

Fig. 14 The architecture of the RNN machine learning of python coding

is 128, and epochs are 10,000. Similarly, a checkpoint function is used to capture the best model. Since the model is trained repeatedly, and the accuracy of the last trained model does not guarantee to be the best one. Most of the times, the model reaches its peak performance, but would be overwritten by the next trained model. The callbacks function can monitor the validation loss of each epoch and then save the model that is more accurate than the previous trained model.

3.2.2 RNN Result

A similar testing procedure as the 1D CNN are applied here for RNN model accuracy checking. The recorded training and validation histories are plotted and shown as Fig. 15, which shows no sign of overfitting or underfitting. The deviation of RNN is relatively greater than the CNN method, and this is why a checkpoint function is necessary in RNN method for capturing the best trained model. The test MSE in this training is 0.059, which indicates the performance of the model is good to predict the toolpath of the sample's layer.

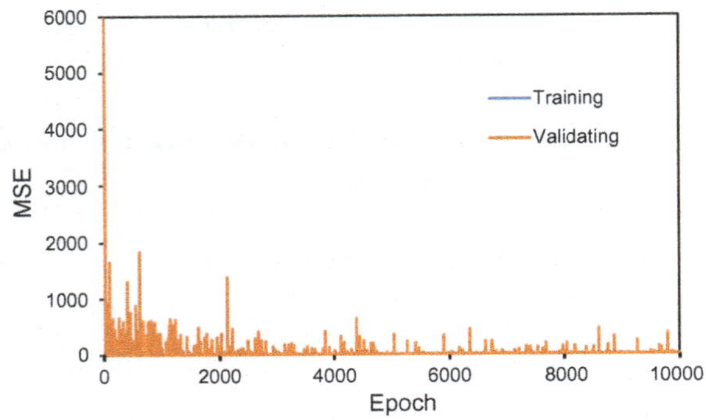

Fig. 15 MSE history of training and testing for each epoch in RNN model

To implement the model, the data prepared for 1D CNN are used. 676 cropped circular images converted with BSIF process, which become 676 rows with 256 features for each image, are saved as CSV files. Then, the CSV file is loaded into Python with the "panda" function. Since the purpose is to predict the angular information for each cropped circular image, 256 features are used as input data and undergo a similar data preparation as the one used for the training dataset, except, there is no output label in these 676 images. After the prediction, a set of values representing the angular information is extracted and saved as a CSV file.

The same method of imposing a direction indicator on the detected fiber direction is used for toolpath reconstruction, shown as Fig. 16. These imposed images are a good local toolpath representation of the tested sample in each sub-sectioned image. It clearly outlines the movement of the printing process, which can be applied to predict the hidden information in a 3D printed sample. Although, only 1 layer is used to demonstrate the toolpath reconstruction, the process can be repeated on the whole sample image stack to acquire the toolpath information of the entire sample. The result can be used as a blueprint for 3D printing reverse engineering or an 3D printing in-situ signature inspection.

4 Summary

The machine learning methods are now widely used in materials design. The opportunities presented by these methods have enabled design of materials with novel properties and reduced time to design complex composite materials for the requirements of specific applications. The present work shows the approaches that can be used for effectively processing the image datasets from materials with the example of a micro-CT scan image dataset processed by three different ML methods. A model composite material specimen is used and the ML approach is used to identify the toolpath used in 3D printing of this specimen. While these methods are useful for a variety of materials related problems such as design of new materials, processing parameter optimization and also defect detection in the microstructure, the ML methods also

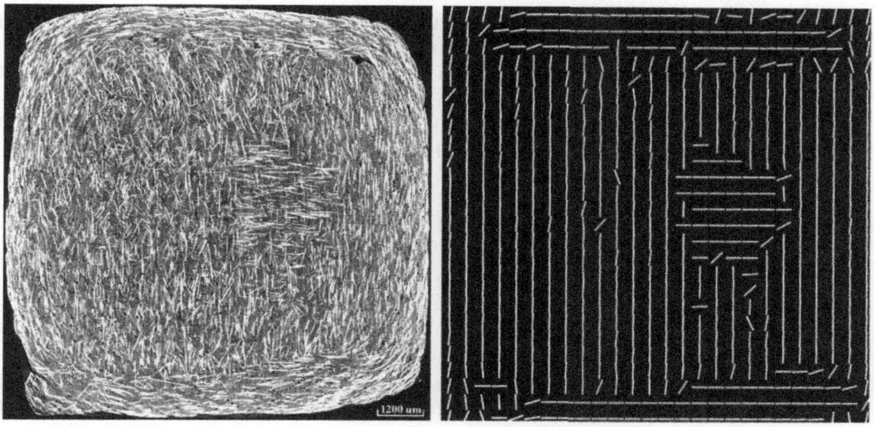

(a) (b)

Fig. 16 **a** The to-be-test CT scan image showing the printing direction by glass fibers and **b** the collection of direction indicators with the trained model showing the toolpath of the certain layer

present challenges that they make the reverse engineering of the products easier. Although the size and geometry of a component can be 3D scanned very easily using available scanners and imaging tools, the quality of a component largely depends on the microstructure. The reverse engineering of microstructure by recovering the toolpath presents a vulnerability that can make reverse engineered products of high quality. The present work shows the need for developing new toolpath methodologies that are difficult to process through ML algorithms.

Acknowledgements National Science Foundation SaTC-EDU grant # 1931724 is acknowledged for supporting this work. The authors thank NYU Tandon School of Engineering Makerspace for the facilities provided for micro-CT scan.

References

1. Kaschel FR, Vijayaraghavan RK, Shmeliov A, McCarthy EK, Canavan M, McNally PJ, Dowling DP, Nicolosi V, Celikin M (2020) Mechanism of stress relaxation and phase transformation in additively manufactured Ti-6Al-4V via in situ high temperature XRD and TEM analyses. Acta Mater 188:720–732
2. Spowart JE, Gupta N, Lehmhus D (2018) Additive manufacturing of composites and complex materials. JOM 70(3):272–274
3. Palmero EM, Casaleiz D, de Vicente J, Hernández-Vicen J, López-Vidal S, Ramiro E, Bollero A (2019) Composites based on metallic particles and tuned filling factor for 3D-printing by fused deposition modeling. Compos Part A Appl Sci Manuf 124:105497
4. Liu Z, Li M, Weng Y, Qian Y, Wong TN, Tan MJ (2020) Modelling and parameter optimization for filament deformation in 3D cementitious material printing using support vector machine. Compos B Eng 193:108018

5. Chen F, Mac G, Gupta N (2017) Security features embedded in computer aided design (CAD) solid models for additive manufacturing. Mater Des 128:182–194
6. Gupta N, Tiwari A, Bukkapatnam STS, Karri R (2020) Additive manufacturing cyber-physical system: supply chain cybersecurity and risks. IEEE Access 8:47322–47333
7. Ngo TD, Kashani A, Imbalzano G, Nguyen KTQ, Hui D (2018) Additive manufacturing (3D printing): a review of materials, methods, applications and challenges. Compos B Eng 143:172–196
8. Justo J, Távara L, García-Guzmán L, París F (2018) Characterization of 3D printed long fibre reinforced composites. Compos Struct 185:537–548
9. Averardi A, Cola C, Zeltmann SE, Gupta N (2020) Effect of particle size distribution on the packing of powder beds: a critical discussion relevant to additive manufacturing. Mater Today Commun 24:100964
10. Wang X, Jiang M, Zhou Z, Gou J, Hui D (2017) 3D printing of polymer matrix composites: a review and prospective. Compos B Eng 110:442–458
11. Heidari-Rarani M, Rafiee-Afarani M, Zahedi AM (2019) Mechanical characterization of FDM 3D printing of continuous carbon fiber reinforced PLA composites. Compos B Eng 175:107147
12. Zeltmann SE, Gupta N, Tsoutsos NG, Maniatakos M, Rajendran J, Karri R (2016) Manufacturing and security challenges in 3D printing. JOM 68(7):1872–1881
13. Yanamandra K, Chen GL, Xu X, Mac G, Gupta N (2020) Reverse engineering of additive manufactured composite part by toolpath reconstruction using imaging and machine learning. Compos Sci Technol 198:108318
14. Bartlett JL, Jarama A, Jones J, Li X (2020) Prediction of microstructural defects in additive manufacturing from powder bed quality using digital image correlation. Mater Sci Eng A 794:140002
15. Huang Z, Dantan J-Y, Etienne A, Rivette M, Bonnet N (2018) Geometrical deviation identification and prediction method for additive manufacturing. Rapid Prototyping J 24(9):1524–1538
16. Kyogoku H, Ikeshoji T-T (2020) A review of metal additive manufacturing technologies: mechanism of defects formation and simulation of melting and solidification phenomena in laser powder bed fusion process. Mech Eng Rev 7(1):19-00182–19-00182
17. Kim C, Espalin D, Cuaron A, Perez MA, MacDonald E, Wicker RB (2018) Unobtrusive in situ diagnostics of filament-fed material extrusion additive manufacturing. IEEE Trans Compon Packag Manuf Technol 8(8):1469–1476
18. Kuipers T, Doubrovski EL, Wu J, Wang CCL (2020) A framework for adaptive width control of dense contour-parallel toolpaths in fused deposition modeling. Comput Aided Des 128:102907
19. du Plessis A, Yadroitsava I, Yadroitsev I (2020) Effects of defects on mechanical properties in metal additive manufacturing: a review focusing on X-ray tomography insights. Mater Des 187:108385
20. Li W, Mac G, Tsoutsos NG, Gupta N, Karri R (2020) Computer aided design (CAD) model search and retrieval using frequency domain file conversion. Addit Manuf 36:101554
21. Comminal R, Serdeczny MP, Pedersen DB, Spangenberg J (2019) Motion planning and numerical simulation of material deposition at corners in extrusion additive manufacturing. Addit Manuf 29:100753
22. Honarvar F, Varvani-Farahani A (2020) A review of ultrasonic testing applications in additive manufacturing: defect evaluation, material characterization, and process control. Ultrasonics 108:106227
23. Deshpande AM, Minai AA, Kumar M (2020) One-shot recognition of manufacturing defects in steel surfaces. Procedia Manuf 48:1064–1071
24. Caggiano A, Zhang J, Alfieri V, Caiazzo F, Gao R, Teti R (2019) Machine learning-based image processing for on-line defect recognition in additive manufacturing. CIRP Ann 68(1):451–454
25. Chen F, Yu JH, Gupta N (2019) Obfuscation of embedded codes in additive manufactured components for product authentication. Adv Eng Mater 21(8):1900146
26. Chen F, Luo Y, Tsoutsos NG, Maniatakos M, Shahin K, Gupta N (2019) Embedding tracking codes in additive manufactured parts for product authentication. Adv Eng Mater 21(4):1800495

27. Chen F, Zabalza J, Murray P, Marshall S, Yu J, Gupta N (2020) Embedded product authentication codes in additive manufactured parts: imaging and image processing for improved scan ability. Addit Manuf 35:101319
28. Sacco C, Baz Radwan A, Anderson A, Harik R, Gregory E (2020) Machine learning in composites manufacturing: a case study of automated fiber placement inspection. Compos Struct 250:112514
29. Xu X, Gupta N (2019) Application of radial basis neural network to transform viscoelastic to elastic properties for materials with multiple thermal transitions. J Mater Sci 54(11):8401–8413
30. Xu X, Gupta N (2019) Artificial neural network approach to determine elastic modulus of carbon fiber-reinforced laminates. JOM 71(11):4015–4023
31. Meng L, McWilliams B, Jarosinski W, Park H-Y, Jung Y-G, Lee J, Zhang J (2020) Machine learning in additive manufacturing: a review. JOM 72(6):2363–2377
32. Liu Y, Guo B, Zou X, Li Y, Shi S (2020) Machine learning assisted materials design and discovery for rechargeable batteries. Energ Storage Mater 31:434–450
33. Khan A, Ko D-K, Lim SC, Kim HS (2019) Structural vibration-based classification and prediction of delamination in smart composite laminates using deep learning neural network. Compos B Eng 161:586–594
34. Dogan A, Birant D (2021) Machine learning and data mining in manufacturing. Expert Syst Appl 166:114060
35. Do DTT, Lee D, Lee J (2019) Material optimization of functionally graded plates using deep neural network and modified symbiotic organisms search for eigenvalue problems. Compos B Eng 159:300–326
36. Xu X, Gupta N (2019) Artificial neural network approach to predict the elastic modulus from dynamic mechanical analysis results. Adv Theor Simul 2(4):1800131
37. El Kadi H (2006) Modeling the mechanical behavior of fiber-reinforced polymeric composite materials using artificial neural networks—a review. Compos Struct 73(1):1–23
38. Sharma A, Kushvaha V (2020) Predictive modelling of fracture behaviour in silica-filled polymer composite subjected to impact with varying loading rates using artificial neural network. Eng Fract Mech 239:107328
39. Kwon O, Kim HG, Ham MJ, Kim W, Kim G-H, Cho J-H, Kim NI, Kim K (2020) A deep neural network for classification of melt-pool images in metal additive manufacturing. J Intell Manuf 31(2):375–386
40. Tian L, Fan Y, Li L, Mousseau N (2020) Identifying flow defects in amorphous alloys using machine learning outlier detection methods. Scripta Mater 186:185–189
41. Ko H, Witherell P, Lu Y, Kim S, Rosen DW (2020) Machine learning and knowledge graph based design rule construction for additive manufacturing. Addit Manuf 101620
42. Wang P, Fan E, Wang P (2020) Comparative analysis of image classification algorithms based on traditional machine learning and deep learning. Pattern Recogn Lett. https://doi.org/10.1016/j.patrec.2020.07.042
43. Xu X, Gupta N (2018) Determining elastic modulus from dynamic mechanical analysis: a general model based on loss modulus data. Materialia 4:221–226
44. Xu X (2020) Machine learning approach to characterize elastic, viscoelastic, relaxation and creep behavior of materials. New York University Tandon School of Engineering, Ann Arbor, p 107
45. Xu X, Elgamal M, Doddamani M, Gupta N (2020) Tailoring composite materials for nonlinear viscoelastic properties using artificial neural networks. J Compos Mater 0021998320973744
46. Sabiston T, Inal K, Lee-Sullivan P (2020) Application of artificial neural networks to predict fibre orientation in long fibre compression moulded composite materials. Compos Sci Technol. http://doi.org/10.1016/j.compscitech.2020.108034
47. Yang J, Yao H (2020) Automated identification and characterization of two-dimensional materials via machine learning-based processing of optical microscope images. Extreme Mech Lett 39:100771
48. Kannala J, Rahtu E (2012) BSIF: binarized statistical image features. In: Proceedings of the 21st international conference on pattern recognition (ICPR2012). Tsukuba International Congress Center Tsukuba Science City, Japan

49. Raschka S, Mirjalili V (2017) Python machine learning. Packt Publishing Ltd., UK
50. Sak H, Senior AW, Beaufays F (2014) Long short-term memory recurrent neural network architectures for large scale acoustic modeling

AI/ML for Quantification and Calibration of Property Uncertainty in Composites

Emil Pitz and Kishore Pochiraju

1 Introduction and Prior Work in Modeling and Uncertainty Quantification

In an engineering context, UQ problems typically consist of a mathematical model representing a physical process [3, 67, 81, 82], such as the growth of damage in a composite structure. Uncertainties can arise from both the correctness of the model itself and the model parameter values [82]. On a general level, UQ requires the characterization of the model parameter uncertainties followed by statistical analysis of the model responses resulting from the forward propagation of the uncertainties through the model [3]. This allows analysis of the induced probability distributions of the model outputs or assessment of the reliability of the modeled system and the prediction quality [82]. Given some observations of the output parameters, inverse UQ allows inference of a statistical distribution of the input parameters [3, 82]. By determining the influence of variability on a model, UQ enables the assessment of the confidence in model predictions.

Of particular interest in composite engineering applications is reducing the required experimental effort in design and certification of parts and structures in aerospace applications, where safety is paramount. Currently, certification of composite components in aerospace is still driven by exhaustive experimental testing, raising product development costs [34]. In 1976 the Federal Aviation Administration (FAA) wrote, "*Past experience with simulation techniques indicate they contain the potential of contributing significantly to the safe and rapid introduction of these new technologies.*" [18], recognizing the importance of simulation in developing and

E. Pitz · K. Pochiraju (✉)
Department of Mechanical Engineering, Stevens Institute of Technology,
Castle Point on Hudson, Hoboken, NJ 07030, USA
e-mail: kpochira@stevens.edu

E. Pitz
e-mail: epitz1@stevens.edu

V. Kushvaha et al. (eds.), *Machine Learning Applied to Composite Materials*,
Composites Science and Technology, https://doi.org/10.1007/978-981-19-6278-3_3

certifying new and increasingly complex airplane systems and structures. Examples of simulations aiding certification in airplane development are airworthiness certification before the first test-flight of a newly developed airframe using simulation [100] or a certified simulation tool for lightning strike analysis and used in the certification of an aircraft (Alenia Aeronautica C-27J) [7]. Hawker de Havilland Aerospace used bird-strike simulations to reduce the required experimental effort and certification cost of a composite trailing edge for the Boeing 787 [35].

Certification by simulation requires validation and verification of the simulation tools. Validation is confirming the suitability of the underlying principles and mathematical models describing physical phenomena. On the other hand, verification examines whether the design tool actually produces the expected results based on the underlying models [34, 69]. Additionally, uncertainties arise from model parameters, such as geometry, loads, boundary conditions, material (strongly influenced by material heterogeneity and manufacturing influences), or environmental influences [3, 78, 81]. These uncertainties are commonly subdivided into aleatoric and epistemic uncertainties [3, 82]. Aleatoric uncertainty is a result of an inherently variable phenomenon, such as the Uncertainty Principle in quantum mechanics [36], while epistemic uncertainty arises from lack of knowledge [3]. In reality, both types of uncertainties are present and sometimes not clearly distinguishable [82]. Using UQ to assess and take into account these uncertainties allows finding an optimum system design in view of uncertainties occurring during manufacturing at the cost of increased computational effort [81]. A purely deterministic analysis cannot be representative of all possible outcomes of an uncertain system [81]. UQ can provide a range of responses in the real world, which can replace some real world testing and, therefore, reduce certification costs compared to the current testing-based approach [34].

The main aim of this work is the development of ML methods to model uncertainties in the material response of composite structures, forward propagate these uncertainties, and assess their impact on structural response. The developed methodologies can be an important step toward certification of aerostructures, materials, and manufacturing processes by simulation.

This book chapter is organized as follows: Sect. 1 gives an introduction to UQ methods, explains the theoretical background of representing stochastic processes, gives an overview of uncertainty propagation methods, presents UQ methods used for composite materials and Machine Learning (ML) methods for UQ, and states the research objective. In Sect. 2, a method to reduce the required samples in uncertainty propagation compared to classical MC method is implemented and presented. Section 3 entails a framework using neural networks to infer random field correlation parameters from experimental strain distributions.

1.1 Representation of Property Variations in Materials

UQ requires the representation of uncertainty using mathematical concepts [81]. Uncertainties, such as spatial variations in material parameters, can be modeled using stochastic processes and fields. When using Finite Element Method (FEM), a stochastic process can be discretized into a denumerable set of random variables [36]. The stochastic process requires the definition of its probability distribution and covariance function, which describes the spatial covariance of the properties at two points [36, 78, 81]. Let X and Y be two jointly distributed random variables with expected values μ_X and μ_Y. Their covariance is given by [74]:

$$\mathrm{Cov}\,(X, Y) = E\left[(X - \mu_X)(Y - \mu_Y)\right], \tag{1}$$

where $E[\;]$ is the expected or average value. If both X and Y tend to be larger than their respective mean, the value of $\mathrm{Cov}\,(X, Y)$ will be positive, while, if X tends to be larger than its mean and Y smaller, the covariance will be negative [74]. Zero covariance indicates independent X and Y. Definition of probability distribution and covariance function is generally a difficult task and commonly assumptions are made instead of inferring from experimental data. Alternative ways of defining the random field will be addressed in this chapter.

Random processes can be represented using spectral representation, which in its most general form is given by [36]:

$$w\,(\boldsymbol{x}, \theta) = \int g\,(\boldsymbol{x})\, d\mu\,(\theta)\,. \tag{2}$$

w is a realization of the random process with the position vector $\boldsymbol{x}$, θ belongs to the space of random events $\boldsymbol{\Omega}$, $g\,(\boldsymbol{x})$ is a deterministic function, and $d\mu\,(\theta)$ is an orthogonal set function. The covariance function $C_{ww}\,(\boldsymbol{x}_1, \boldsymbol{x}_2)$ of the stochastic process for the position vectors $\boldsymbol{x}_1$ and $\boldsymbol{x}_2$ can be decomposed:

$$C_{ww}\,(\boldsymbol{x}_1, \boldsymbol{x}_2) = \int g\,(\boldsymbol{x}_1)\, g\,(\boldsymbol{x}_2)\, d\mu_1\,(\theta)\, d\mu_2\,(\theta)\,. \tag{3}$$

A commonly used alternative formulation to the spectral representation is Karhunen-Loève Expansion (KLE), where the random process is represented in terms of a denumerable set of orthogonal random variables in a Fourier-type series [36, 81]:

$$w\,(\boldsymbol{x}, \theta) = \sum_{n=0}^{\infty} \sqrt{\lambda_n}\, \xi_n\,(\theta)\, f_n\,(\boldsymbol{x})\,, \tag{4}$$

with random variables $\xi_n\,(\theta)$, constants λ_n, and orthonormal functions $f_n\,(\boldsymbol{x})$. At this place Polynomial Chaos Expansion (PCE) should also be mentioned as an alternative and commonly used representation of a stochastic process. PCE uses a linear combination of orthogonal Hermite (in the case of a Gaussian process) polynomials to represent the random process [81, 88, 92].

If the mean and covariance function of a random process are constant over space, the random field is called stationary [4]. Furthermore, if the covariance function is only dependent on the Euclidian distance between two points, the process is isotropic [4]. An example for non-stationary random events is the precipitation in the United States with strongly varying average values in different regions [32]. Using KLE, both Gaussian and non-Gaussian, as well as stationary and non-stationary, and isotropic and non-isotropic random processes can be modeled [63].

1.2 Sampling with Monte Carlo Methods

One main objective of UQ is to forward propagate the uncertainties of the input parameters through the model and analyze the impact of these uncertainties on the model response. One of the simplest and most widely used methods is the MC method. MC entails drawing samples of input parameters from a target distribution, evaluating the model deterministically using the individual samples and recording each model response to perform statistics [37, 47]. It should be noted that this requires the description of the input variables' probability distributions, which can be demanding depending on available data. Evaluating a model given by $f(x)$ with input x, using n samples $U_1, U_2, \ldots, U_n$ randomly drawn from the respective distribution, gives the MC-estimate of the mean of the model [28, 37]:

$$\hat{\alpha}_n = E\left[f(x)\right] = \frac{1}{n}\sum_{i=1}^{n} f(U_i). \tag{5}$$

When the material parameters in FEM analyses are represented using random processes, MC can be applied to examine the response statistics. A schematic of MC method in the context of FEM analysis is given in Fig. 1.

According to the Central Limit Theorem, the error of the MC-estimate follows a normal distribution with mean 0 and standard deviation $\frac{\sigma_f}{\sqrt{n}}$, where σ_f is the standard deviation of the function $f(x)$ [37, 75]. That means, if the expected standard error of the MC-estimate should be halved, the number of sample points has to be quadrupled. As σ_f is usually not known, the sample standard deviation can be used to estimate the error [37]:

$$\sigma_f \approx s_f = \sqrt{\frac{1}{n-1}\sum_{i=1}^{n}\left[f(U_i) - \hat{\alpha}_n\right]^2}. \tag{6}$$

The main advantage of MC method is the fact that the standard error is independent of the dimensionality of f [75]. However, with the expected error being $\mathcal{O}\left(n^{-1/2}\right)$, a large number of model evaluations might be required to achieve sufficient levels of accuracy. In case of computationally expensive model evaluations, such as progressive damage simulations of composite structures, which can take hours to days for a single evaluation, MC method might become prohibitive. A method to improve the convergence behavior of MC by choosing samples deterministically rather than randomly from a stochastic distribution is the main subject of Sect. 2.

1.3 Sampling with Reliability Methods

While MC method enables extensive statistical analysis on the model output parameters, its computational ineffectiveness can be prohibitive and distribution fitting might not always be required. In a number of applications computationally less demanding reliability methods can be an attractive alternative. Reliability methods require a scalar response function $g(\boldsymbol{x})$, where $\boldsymbol{x} = (x_1, x_2, \ldots x_n)$ are uncertain input variables [3, 24]. $g(\boldsymbol{x}) < 0$ implies failure, while $g(\boldsymbol{x}) > 0$ constitutes no failure [24]. The fundamental objective of reliability methods is the calculation of the failure probability $P_f = P[g(\boldsymbol{x}) \leq 0]$ based on probabilistic approaches. This could for example be the failure probability of a structural component [3, 24]. The main advantage of reliability methods compared to MC is improved computational efficiency as fewer model evaluations are required. At this place a more detailed introduction to the commonly used First-Order Reliability Method (FORM) will be given and the Second-Order Reliability Method (SORM) will be introduced. A number of additional reliability methods can be found in literature [3].

Using the probability densities $f_i(x_i)$ of the random input variables, which are assumed to be independent, the failure probability can be calculated as [24, 36]:

$$P_f = \int_F f_1(x_1) \ldots f_n(x_n)\, dx_1 \ldots dx_n, \tag{7}$$

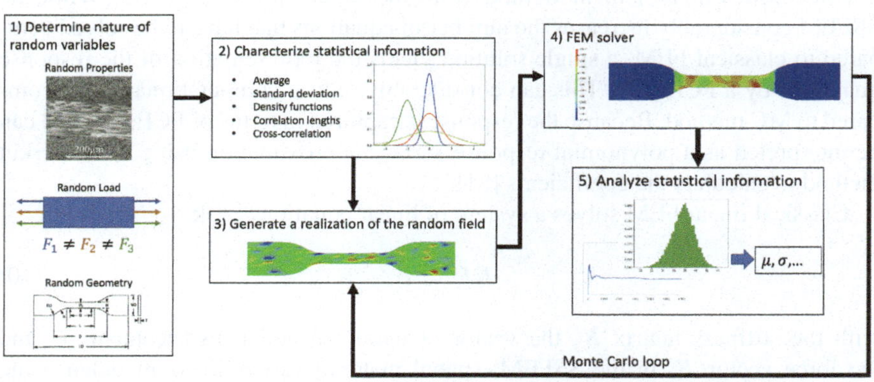

Fig. 1 Schematic representation of MC technique using FEM analysis (adapted from [8])

where F is the failure domain. For standard normally distributed input variables, P_f can be calculated using the cumulative standard normal distribution Φ [24]:

$$P_f = \Phi(-\beta), \tag{8}$$

with the reliability index β. The reliability index gives the distance between the origin of the design space and the so called design point or most likely failure point $\boldsymbol{x}^*$ on the failure surface and can be calculated using [24]:

$$\beta = \frac{g(x^*) - \boldsymbol{a} \cdot \boldsymbol{x}}{|\boldsymbol{a}|}, \qquad \boldsymbol{a} = \nabla g(\boldsymbol{x}). \tag{9}$$

The design point contributes most to the failure probability and is the point on the failure surface with the minimum distance to the origin of the space of the input variables [24]. This point is iteratively determined. If $g(\boldsymbol{x})$ is a non-linear function, the function can be linearized at the failure point and Eq. 8 gives an approximate failure probability. Furthermore, if the input parameters are not standard normally distributed, they have to be transformed to the standard normal space.

While FORM uses a plane in the input variables' space to approximate the failure surface at the design point, SORM uses a quadratic surface [24]. This can improved accuracy of the predicted failure probability. Examples for applications of FORM and SORM can be found in [23, 87] for the analysis of a cracked steam generator tubing and aerospace applications.

1.4 Sampling with Spectral Stochastic Finite Element Methods

Another method that has attracted considerable interest in recent years is the Spectral Stochastic Finite Element Method (SSFEM) initially proposed in [36]. While the SSFEM considerably increases the number of equations that have to be solved compared to classical FEM, a single solution yields the representation of the response quantities by a PCE [81]. This can considerably reduce computational effort compared to MC method. Because the response is captured in terms of PCE, SSFEM can be interpreted as a polynomial response surface approximation using the Galerkin method to calculate the coefficients [81].

Classical linear FEM solves a system of linear equations [20]:

$$\boldsymbol{KU} = \boldsymbol{F}, \tag{10}$$

with the stiffness matrix $\boldsymbol{K}$, the vector of unknown nodal displacements $\boldsymbol{U}$ and the force vector $\boldsymbol{F}$. In the SSFEM, spatial material variability is modeled using KLE representing a Gaussian stochastic field [36, 81]. This results in a stochastic element stiffness matrix. Furthermore, the vector of now random unknown nodal displacements is expanded using PCE. The equilibrium equation of the SSFEM can now be written as:

$$\left(\sum_{i=0}^{\infty} \boldsymbol{K}_i \xi_i(\theta)\right) \cdot \left(\sum_{j=0}^{\infty} \boldsymbol{U}_j \boldsymbol{\Psi}_j(\theta)\right) - \boldsymbol{F} = \boldsymbol{0}, \tag{11}$$

with deterministic matrices $\boldsymbol{K}_i$ corresponding to the terms of the KLE of the global random stiffness matrix, random variables $\xi_i(\theta)$, and deterministic vectors $\boldsymbol{U}_j$ and random Hermite polynomials $\boldsymbol{\Psi}_j$ from PCE of the random nodal displacement vector. Keeping only a finite number of terms P in the PCE and after some algebraic manipulations, Eq. 11 can be rewritten as a system of linear equations of size $NP \times NP$, where N, in the 1-dimensional case, is the number of nodes in the model :

$$\mathcal{K}\mathcal{U} = \mathcal{F}. \tag{12}$$

The size of the system of linear equations in deterministic FEM would only be $N \times N$ for the same number of nodes.

SSFEM is exact for an infinite number of retained terms [81]. To reduce the computational effort however, a large number of terms can be prohibitive. This requires the correlation length of the random field represented by KLE to be sufficiently large for the chosen number of retained terms, to obtain a good approximation. While SSFEM can yield results with lower computational effort compared to MC method, its application is practically limited to linear or very simple non-linear problems [81] (e.g., elasto-plastic problem [6]).

As an alternative to the SSFEM, perturbation based Stochastic Finite Element Method (SFEM) shall be mentioned as well. The Perturbation approach uses a Tailor-series approximation of the stochastic stiffness matrix, force vector, and displacement vector [50, 81]. In practice, higher than first-order approximations of the perturbation approach are only feasible for Gaussian random fields [81]. Some recent applications to dynamical systems [31, 50] and geometrically non-linear problems [44] can be found in literature.

1.5 Approaches for Micro- and Macro-UQ of Composite Performance

While uncertainties can arise from multiple model inputs, such as variability in the applied load, in boundary conditions, or part geometry, studies in literature predominantly deal with uncertainty of the composite material response. Due to their heterogeneity and strong dependency of the material microstructure on manufacturing influences [95], composite materials can show a larger degree of uncertainty compared to homogeneous materials [97]. A main difference between studies dealing with composite material uncertainty is the use of either random variables (e.g., [5]), where the material parameter is random but constant in the entire domain, or random fields (e.g., [80]) with spatially varying material parameters. One of the main challenges in UQ of composite structural response is the determination of the distributions of random variables or distributions and correlation behavior of random fields [21]. Because of complex material behavior and usually inadequate size of experimental datasets, assumptions have to be made about the random fields governing uncertain material parameters [21, 95]. This leads multiple studies to conclude that micromechanics models, as means of virtual testing frameworks, are required to correctly determine the random fields [21, 95].

Generally, studies for UQ in composites can be distinguished into two main approaches: Micromechanics based methods that infer the stochastic material response from the material micro- and/or mesostructure, and studies dealing with stochastic macro analysis of composites [78].

Multiple different representations of the material heterogeneity have been used in micromechanics-based studies. A preferred method is the FEM simulation of Representative Volume Element (RVEs) or Stochastic Volume Element (SVEs) of the composite microstructure with random placement of fibers (e.g., [95]), random fiber waviness (e.g., [43]), or random constituent behavior (e.g., [10]). Randomness in fiber placement or fiber waviness is commonly characterized from micrographs [84, 85, 94, 95], while other studies directly use scatter from experimental data [10]. Woven or braided composites are often modeled using a multi-scale approach, where uncertainties are propagated from the micro-scale (fibers and matrix) to the meso-scale (tow geometry) and the macro-scale (coupon level) [11, 15, 85]. Because solving RVEs or SVEs using FEM is generally computationally expensive, some studies have used stochastic RVEs or SVEs to train surrogate models or fit Reduced Order Models (ROMs) [10, 11, 43, 95].

Instead of using RVEs or SVEs, some studies have used Mean-Field Homogenization (MFH) approaches (e.g., [62] for Short-Fiber Reinforced Composite (SFRPs), where Bayesian Interference (BI) was used to identify model parameters) or purely analytical models (e.g., [5]) to estimate the random composite response based on variability in the constituents.

The second class of UQ-approaches for composites directly models the macro-scale (homogenized) composite response without taking into account the constituents' behavior or distribution. While a number of studies have used random variables to model effective composite properties (e.g., [48, 57]), at this place only studies that used random fields will be covered as these are of more interest for this chapter. Furthermore, multiple studies have used spatial random fields to model material parameters with purely estimated random field parameters or varying random field parameters to examine their influence on structural response (e.g., [64, 65, 97, 103]).

Of particular interest for this chapter is a stochastic hyperelastic anisotropic material model for composites presented in [80]. An inverse method was used to determine the random field correlation lengths and variation from experimental Digital Image Correlation (DIC) strain distributions based on a maximum likelihood estimation. A similar approach was used in [22] to model the stochastic response of additively manufactured materials. The authors used a maximum likelihood method to fit model parameters from experimental response curves, though mention that a least-square method or a Bayesian approach could have been used as well. The correlation length was not fitted to experimental data as only the global response was used. However, the authors say, given full-field measurements, this could have been inferred as well.

1.6 Machine Learning Methods for UQ

With ever increasing popularity due to their versatility and advances in computational resources, ML techniques have also increasingly been applied to UQ problems. As UQ often requires a large number of model evaluations and computational efficiency is required, frequently ML methods are used to create computationally efficient surrogate models, which are evaluated instead of or in addition to a high fidelity model. Most of these methods have in common that the surrogate model is used in a deterministic sense, similar to a deterministic simulation where samples are drawn randomly from a distribution for evaluation. Uncertainties are not embedded in the neural network surrogates. An example is the use of a neural network as a surrogate model for low fidelity simulations combined with a high fidelity FEM model to perform UQ in woven composites in [11]. In [39], a neural network surrogate was employed to perform MC on Molecular Dynamics (MD) simulations of graphene with defects. To increase the efficiency of MC method, in [59] a neural network was used to evaluate low cost samples to find points enabling efficient sampling of the distribution. Points that were found to efficiently sample the space were then used in a high fidelity model to perform MC on groundwater flow.

To directly embed uncertainty into a neural network, some studies have used Bayesian Neural Networks (BNNs). Instead of using deterministic parameters as is the case in classical neural networks, in BNNs the weights are represented by distributions [26]. This enables the BNN to consider uncertainties in both the data and the model itself [71]. A schematic comparison of a classical neural network and a BNN is given in Fig. 2. In [71] a BNN was used to model the stochastic response of composite SVEs with plastic matrix response including both aleatoric (mainly a result of random fiber placement in the SVE) and epistemic (result of small datasets) uncertainty. Once the BNN was trained, the effective composite response could be directly estimated including uncertainties. A similar approach was used in [26] to performing time-dependent reliability analysis of structures.

A small number of studies have used ML techniques to extract stochastic microstructure-property relationships. Using Microcomputed Tomography (micro-CT) images of woven composites, autocovariance functions (two-point statistics) were fitted to the microstructure in [70]. This was followed by dimensionality reduction and training of a linear regression model to predict the elastic properties based on the autocovariance function. To create stochastic RVEs, in [14] digitized microstructure images for multiple different inclusions (porosity, isotropic, anisotropic) were used to train a classification tree. The trained model was then used to create new microstructures exhibiting the same stochastic features as the training data.

2 Enhancing Sample Efficiency Through Quasi Monte Carlo Simulations

Monte Carlo (MC) technique is commonly used in engineering applications for UQ due to its straightforward and versatile applicability [40]. MC entailed randomly drawing several samples for the model parameters from known probability distributions. The underlying deterministic model has to be executed for each instance drawn from the distribution, and the model responses are tabulated [40, 98]. The output metrics (model responses) are characterized, and appropriate statistical estimates (mean, standard deviation, mode) for the response's statistical representation are computed.

One significant advantage of MC is that the expected error of the estimates is $\mathcal{O}\left(n^{-1/2}\right)$, where n is the number of model evaluations. It is independent of the dimensionality of the model [75]. However, MC can be practically infeasible in computationally expensive applications as its convergence will require numerous simulations. Based on the central limit theorem and the continuous mapping theorem, the standard error of the mean (ϵ_n) of a MC simulation can be used as a measure of error [47]:

$$\epsilon_n = \frac{\hat{\sigma}_n}{\sqrt{n}} \tag{13}$$

where the variance of the random variate is estimated using the variance of the simulations ($\hat{\sigma}_n$), with x_i being the result of a single random simulation [47]:

$$\hat{\sigma}_n = \sqrt{\left(\frac{1}{n}\sum_{i=1}^{n} x_i^2\right) - \left(\frac{1}{n}\sum_{i=1}^{n} x_i\right)^2} \tag{14}$$

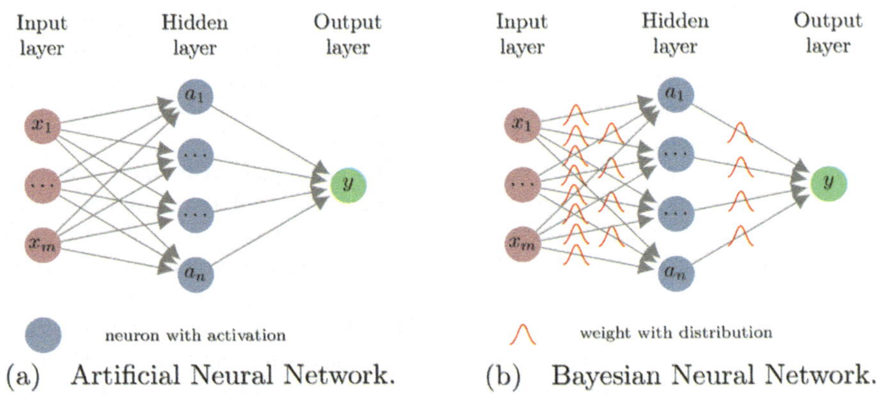

Fig. 2 Conventional neural network with deterministic weights (**a**) and BNN with weights represented by distributions (**b**) (adapted from [26])

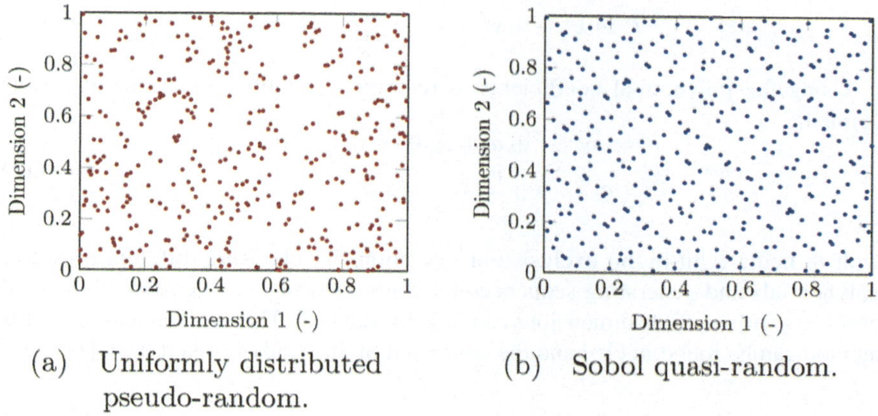

Fig. 3 Sample point distributions of **a** uniformly distributed pseudo-random points and quasi-random, **b** Sobol sequence exhibiting lower discrepancy (reprinted from [73])

QMC technique can reduce the number of samples required to obtain sufficient convergence. Instead of drawing samples from a random distribution, samples are derived deterministically from Low Discrepancy Sequence (LDSs) [75]. Discrepancy describes the non-uniformity of points in a set. Low discrepancy samplings better 'fill' space without exhibiting 'clumps' or 'empty holes' in space. Examples of LDSs are Halton, Faure, Niederreiter, or Sobol sequences [47, 68, 77]. Figure 3 depicts the sample distributions obtained by a uniformly distributed pseudo-random sampler and the Sobol quasi-random sample in a 2D space. The Sobol LDS distributes point samples more uniformly (Fig. 3b) across the sample space.

Another commonly used method to reduce the number of required samples compared to MC is Latin Hypercube Sampling (LHS) [76]. LHS provides sample points that are well spread out over the unit cube and, therefore, can reduce variation in the sampled function [76]. However, multiple studies have shown improved convergence behavior for QMC using LDSs [49, 54, 76, 90, 91] compared to LHS. Furthermore, the superiority of QMC with Sobol sequences over LHS has been shown especially for higher-dimensional problems [54, 76]. Since QMC with a Sobol sequence has shown improved convergence in multiple studies and for different applications, we used a Sobol sequence [77] for spatially mapping strength distributions in this study.

Obtaining a one dimensional Sobol sequence $x^1, x^2, \ldots, 0 < x^i < 1$ over a unit interval requires a set of direction numbers v_i [16]:

$$v_i = \frac{m_i}{2^i} \tag{15}$$

where m_i is an odd integer, $0 < m_i < 2^i$. Generally v_i are derived from the coefficients of an arbitrarily chosen primitive polynomial P of degree d (coefficients a from $\{0, 1\}$):

$$P \equiv x^d + a_1 x^{d-1} + \cdots + a_{d-1} x + 1. \quad (16)$$

Using the polynomial coefficients, a recurrence relation for calculating v_i is defined:

$$\begin{aligned} v_i &= a_1 v_{i-1} \oplus a_2 v_{i-2} \oplus \cdots \oplus a_{d-1} v_{i-d+1} \\ &\oplus v_{i-d} \oplus \frac{v_{i-d}}{2^d}, \quad i > d, \end{aligned} \quad (17)$$

with $\oplus$ being a bit-by-bit exclusive-or operation. By choosing different primitive polynomials and generating sequences for every coordinate, extensions to general problems with multiple dimensions can be achieved [75]. The implementation of this method can be found in [16] and the code used in this study is taken from [17].

2.1 Application of QMC for the Modeling of Random Fields with KLE

Generation of a realization of a random field using KLE requires m uncorrelated and centered random variables ξ_i (see Eq. 19) [36], where m is the number of terms used in the approximation. To perform QMC, each of the random variables ξ_i was treated as a dimension of a m-dimensional Sobol sequence. Each Sobol number was then inverted using the inverse cumulative normal distribution function (Intel MKL *vdcdfnorminv* function in Fortran [45]) to obtain a Gaussian quasi-random variable.

We now describe the results from testing the hypothesis that the QMC method with multi-dimensional low discrepancy sampling converges faster than classical MC method. Faster convergence shows as smaller variation between the estimates with increasing number of simulations. To test this hypothesis, we performed 10,000 simulations of additively manufactured specimens, respectively, using MC and QMC methods. An elasto-plastic material model was used to model the material response and stiffness and strength were modeled using a random field (cross-correlation between stiffness and strength was one). Figure 4 shows the floating average of the maximum stress (after reaching the maximum stress the material exhibits perfectly plastic behavior and accumulates plastic strain at constant stress) during the simulations and the standard deviation ($\hat{\sigma}_n$) of the estimated maximum stress mean for both MC and QMC. Equation (18) defines the standard deviation of the predictions, $\hat{\sigma}_n$, after n simulations where (x_j) is the result of a single random simulation.

$$\hat{\sigma}_n^2 = \frac{1}{n}\sum_{i=1}^{n}\left(\frac{1}{i}\sum_{j=1}^{i} x_j\right)^2 - \left[\frac{1}{n}\sum_{i=1}^{n}\left(\frac{1}{i}\sum_{j=1}^{i} x_j\right)\right]^2 \quad (18)$$

Faster decreasing standard deviation of the mean estimate relates to faster convergence.

Similar average values of the maximum stress are predicted using MC (37.66 MPa) and QMC (37.70 MPa), with generally more stable predictions from QMC (see Fig. 4). Comparing the standard deviation of the mean estimates, again more stable pre-

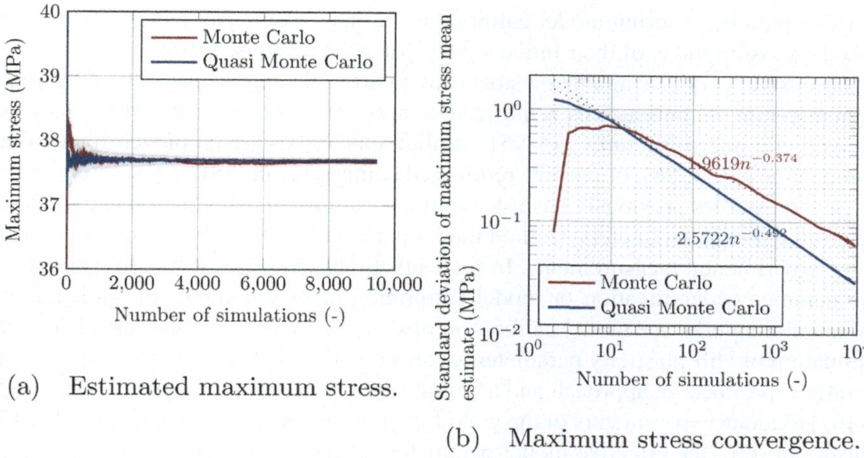

Fig. 4 The shaded intervals represent the 95% confidence interval

dictions and faster convergence are seen using QMC (see Fig. 4). To quantify the improvement in convergence, a power law (an^{α}) fit was made to the standard error data. Coefficients (a) and the powers (α) that relate the number of simulations performed (n) to the standard deviation are shown in Fig. 4. The power law fit for MC gives an exponent of -0.374 while the QMC fit has an exponent of -0.492. This confirms faster convergence of QMC method and shows that QMC method can be used to speed-up convergence when using KLE to approximate spatial random fields.

As an alternative to MC, several studies use bootstrapping methods to reduce the computational burden. Uncertainty quantification in composite stochastic damage analysis has been performed using small sample sizes and bootstrapping techniques [103, 104]. Bootstrapping entails randomly resampling and replacing an initial population of samples (termed as non-parametric bootstrap) and performing statistical analysis on the resampled points [41, 102]. This technique can determine the accuracy of the initial sample set, but it does not improve the parameter estimates (e.g. of the mean) from the sample [41]. The distribution of the bootstrapped samples is always centered at the observed statistic of the initial sample set, not on the population statistic (e.g. mean of the initial sample, not the population mean) [41]. Since bootstrapping does not improve the parameter estimate, more samples will be necessary to increase the accuracy of the parameter estimate.

3 UQ Parameter Calibration with Neural Networks

Identification of random field parameters that describe the spatial variability of the material response is a demanding and difficult task. While random fields have been more commonly used in material modeling applications, they often require computationally expensive microstructure analysis [93–95] or inverse optimization methods

[80] for parameter fitting/model calibration. In other cases, parameter selection is based on assumptions of their influence on structural response [103].

Extensive work is available on stochastic multi-scale approaches, where the random response of continuously reinforced composites is estimated based on analysis of synthetic microstructures [93–95]. Similar approaches are employed for woven composites in [83, 86], or particle reinforced composites in [38].

Several studies previously have shown the feasibility of using neural networks to determine model parameters. Certain model parameters cannot be directly obtained from experimental measurements. In such situations, neural networks [61] simplify the parameter identification or model calibration tasks. Examples of applications of neural networks are model calibration of damage models in sheet metals [1] or estimation of Hill-plasticity parameters in sheet metals [19]. The performance of an iterative optimization approach and a neural network-based approach are compared in [61] to identify parameters of the yield function for additively manufactured ABS plastic. In [66], the effective model parameters of a machine tool drive system are inferred using neural networks and in [55] a CNN is used for identification of material porosity and tortuosity in water-saturated porous materials.

For the present chapter a methodology is developed for modeling additively manufactured structures using Gaussian random fields for strength and stiffness distributions. A CNN is used to extract random field parameters from DIC images, more particular the correlation lengths of the stochastic fields. CNNs are a type of Neural Network (NNW), using convolutions in place of matrix multiplications in order to reduce the number of parameters [55] and are commonly used for image classification or regression tasks [2, 30, 53, 60, 101]. In contrast to inverse methods, once trained, NNWs allow efficient predictions without evaluating a forward model [55]. A large range of engineering applications for NNWs can be found, from model parameter estimation [55] similar to the current problem to surrogate models for predicting the nonlinear response of composite materials [96].

To fully define a Gaussian random field, the mean, standard deviation, and a covariance function are required. The chosen exponential covariance function is this work is defined by two correlation lengths. The correlation lengths define the correlation between any two points in the random field. Mean and standard deviation for stiffness and strength are readily available from experimental tensile tests. A process for determining the correlation lengths using strain distributions from DIC images from experimental tensile tests is presented. The process consists of two sub problems [55]: A forward model for simulating additively manufactured specimens using the correlation lengths and material parameters as input, and the inverse problem to predict the correlation lengths from experimental strain distribution.

To solve the inverse problem, a CNN is developed and trained using strain distributions obtained from the forward model (FEM simulation) with random correlation lengths as input. After training, the strain distributions from DIC measurements are fed into the CNN to predict the correlation lengths for the material of interest.

Once the correlation lengths are determined using the CNN, the random fields are fully defined and the forward model can be used to perform MC simulations for examining the stochastic model response [81].

3.1 Karhunen-Loève Expansion

UQ of stochastic engineering systems requires the correct description of the randomness of model parameters [89]. Spatial random fields can be used to represent spatially varying model parameters [89]. The auto-correlation function of an individual parameter describes the correlation between two points in space [36, 105]. On the other hand, cross-correlation describes similarities between different random parameters [105].

Instances of random fields can be generated using KLE. KLE can be regarded as an abstract discretization of a random process using series expansion of a set of orthogonal random variables, based on modal decomposition of the auto-covariance kernel [33, 36, 52, 103]. An advantage of KLE is that it has the property of minimizing the mean-square error of the finite representation of a random process [36]. The KLE of a random process $w(\boldsymbol{x}, \theta)$ can be written as a truncated sum of m elements [36]:

$$w(\boldsymbol{x}, \theta) = \overline{w}(\boldsymbol{x}) + \sum_{i=0}^{m} \xi_i(\theta) \sqrt{\lambda_i} f_i(\boldsymbol{x}). \tag{19}$$

$\boldsymbol{x}$ is the position vector in the domain $\boldsymbol{D}$, θ belongs to the space of random events $\theta \in \Omega$, $C(\boldsymbol{x}_1, \boldsymbol{x}_2)$ denotes the covariance function with the two position vectors $\boldsymbol{x}_1$ and $\boldsymbol{x}_2$, $\overline{w}(\boldsymbol{x})$ is the expected value of $w(\boldsymbol{x}, \theta)$, and ξ_i are centered and uncorrelated random variables. λ_i and $f_i(\boldsymbol{x})$ can be found as solution to the integral equation [36]:

$$\int_{\boldsymbol{D}} C(\boldsymbol{x}_1, \boldsymbol{x}_2) f_i(\boldsymbol{x}_1) d\boldsymbol{x}_1 = \lambda_i f_i(\boldsymbol{x}_2). \tag{20}$$

We used an absolute exponential kernel to represent the covariance function for a two-dimensional Gaussian random field [36, 103]:

$$C(\boldsymbol{x}_1, \boldsymbol{x}_2) = \sigma^2 \exp\left(-\frac{|x_1 - x_2|}{l_1} - \frac{|y_1 - y_2|}{l_2}\right), \tag{21}$$

where the parameters l_1 and l_2 govern the rate in x and y direction, respectively, at which the correlation between two points $\boldsymbol{x}_1$ and $\boldsymbol{x}_2$ in the domain decays and are commonly called correlation lengths. σ^2 is the variance. It should be noted that this covariance function was a modeling assumption and was mainly chosen for ease of implementation and the ability to represent non-isotropic covariance (by choosing different l_1 and l_2). It is possible that a different covariance function would better fit the encountered data.

3.2 Computational Model

A nonlinear FEM code was developed in-house to efficiently model stochastic problems. The code is written in Fortran and is based on the Intel MKL Pardiso direct solver. OpenMP is used to allow multiple simulations to be run in parallel, which enables a large number of simulations to be solved in a short time. On a standard workstation with an 8 core, 3.20 GHz, AMD Ryzen 7 2700 processor, 100,000 simulations required for training of the NNW were performed in less than two days.

The FEM code uses a small strain formulation and an elasto-plastic material model with incremental J2 plasticity and isotropic hardening [13], using ascending pairs of plastic strain and corresponding stress as input to define the yield surfaces. The code is used for modeling additively manufactured specimens. The random fields for stiffness and yield stress distributions are discretized using two-dimensional KLE for in-plane (xy-plane) variations of the material parameters. Through the thickness (z-direction) the material parameters are held constant. Correlation between points in the two-dimensional random field is governed by two correlation lengths, one for each spatial direction. The computational implementation of KLE is based on the Matlab toolbox FERUM, developed at University of California at Berkeley [27]. Running a FEM simulation, the KLE is evaluated at each integration point.

In Fig. 5 three realizations of a random Gaussian field using KLE are depicted with different correlation lengths, where l_1 is the correlation length of the random field in spatial x-direction and l_2 in y-direction. It is supposed that the correlation length is a result of the material microstructure resulting from the additive manufacturing process and hence the correlation lengths for the stiffness field are similar to the correlation lengths of the strength field.

3.3 Parameter Estimation Using a Neural Network

The objective of this research is to estimate parameters describing a spatial random field based on DIC images by using a NNW, which is trained on data obtained from FEM simulations. A brief introduction to NNWs is given in the following section.

Artificial NNWs can be viewed as function $F : \boldsymbol{x} \rightarrow \boldsymbol{y}$ with input $\boldsymbol{x} = [x_1\ x_2\ \dots\ x_m]^T$ and the output $\boldsymbol{y} = [y_1\ y_2\ \dots\ y_m]^T$ [96]. A NNW is built from artificial neurons, performing a weighted sum operation on the input [55, 96]. In Fig. 6 a neural network with a single hidden layer is exemplary depicted, with each dot representing a neuron and each arrow a connection between two neurons [55]. The network has an input layer with m inputs $x_1, \dots, x_m$, a hidden layer with n output activations $a_1, \dots, a_n$, and an output layer with i outputs $y_1, \dots, y_i$. The activations $a_1, \dots, a_n$ are given with the inputs $x_1, \dots, x_m$ by Refs. [55, 96]:

$$a_j = \sigma\left(\sum_{k=1}^{m} w_{jk}^{(1)} x_k + b_j^{(1)}\right) \quad j = 1, \dots, n \tag{22}$$

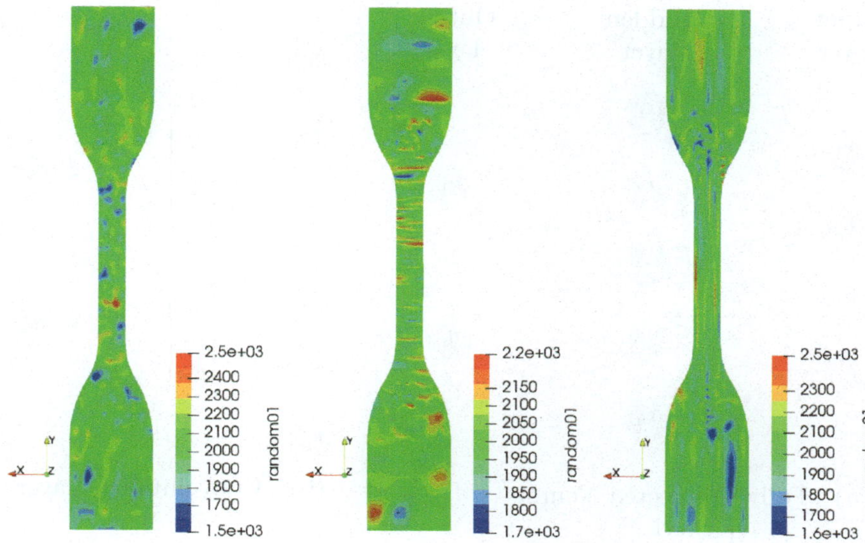

(a) $l_1 = 1, l_2 = 1.$ (b) $l_1 = 1, l_2 = 0.1.$ (c) $l_1 = 0.1, l_2 = 5.$

Fig. 5 Three realizations of the random stiffness field at different correlation lengths [73]

with $w_{jk}^{(1)}$ a weight associated with the respective neuron connection, a bias term $b_j^{(1)}$, and the non-linear activation function σ. The weights w_{jk} are obtained through training of the NNW [96]. Due to fast training performance the Rectified Linear Unit (ReLu) function $\sigma(x) = \max(0, x)$ is a commonly used activation function [53, 55, 56]. Similarly to the activations $a_1, \ldots, a_n$, the outputs $y_1, \ldots, y_i$ of the network are calculated having as input the activations from the hidden layer.

The disadvantage of fully connected NNWs as described above is the possibly enormous number of free parameters [12, 55]. For overcoming this problem, CNNs can be used, where the matrix multiplication in Eq. 22 is replaced with a convolution with a set of filter weights $w_1, \ldots, w_n$ to be trained [55, 56]. A convolutional layer is schematically depicted in Fig. 7. CNNs are designed to process data supplied in the form of multiple arrays, such as colour images, which consist of three 2D arrays containing pixel intensities of the three colour channels [25, 56]. Typically each convolutional layer contains multiple convolutional masks that are applied simultaneously for better performance [55].

For downsampling, that is reducing the dimension of activations, commonly additional pooling layers are used in CNNs [53, 55]. By normalizing layer inputs using batch normalization layers, acceleration of the training can be achieved [46]. NNWs are prone to overfitting, which is shown by good performance for the training dataset but poor performance for unknown datasets [55]. Figuratively speaking, overfitting occurs when a NNW adapts only to the training samples and is unable to generalize [55]. Overfitting can be reduced by using dropout layers [79]. During training,

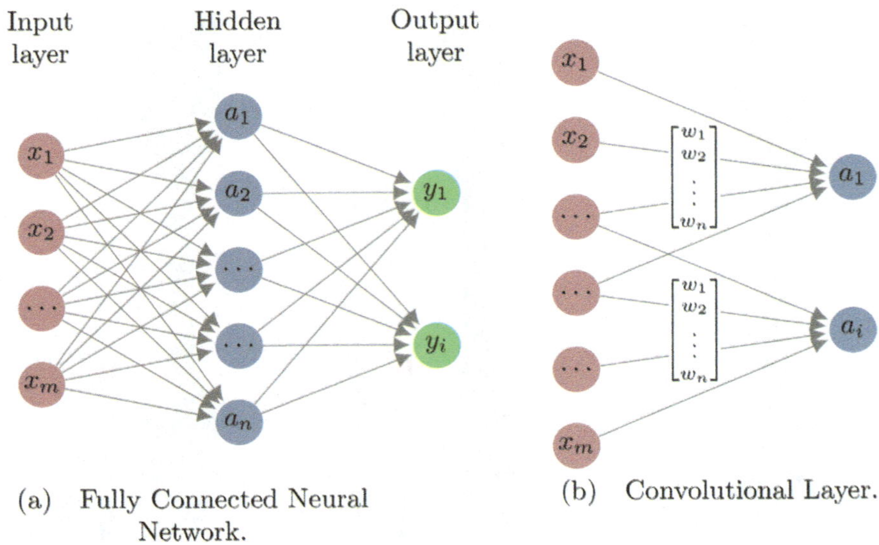

Fig. 6 Fully connected NNW with a single hidden layer (**a**), single convolutional layer (**b**) (adapted from [55])

dropout layers randomly drop units and their corresponding connections from the NNW, preventing units to co-adapt too much [79].

3.4 Network Architecture

The utilized network architecture for this chapter is roughly based on the AlexNet classification network [53]. Instead of the final softmax layer used in the AlexNet network for classification, in the current problem a fully connected layer with linear activation and a single output is used for each correlation length to perform the regression task [30, 60]. The network consists of five convolutional layers with batch normalization before the ReLu activations in each layer and maximum pooling in the second and fifth convolutional layers, respectively. Compared to the AlexNet network the number of filters in the convolutional layers is smaller, which reduces the number of unknowns. After the convolutional layers, the input is flattened and two fully connected layers with batch normalization and dropout are used. These layers are then connected to one fully connected layer for each correlation length, respectively. Using one output layer with one node for each correlation length prediction, respectively, allows evaluation of training parameters for each correlation length individually. A schematic overview of the network architecture is given in Fig. 7 and a more detailed description of the individual layers with input and output parameters can be found in Table 1. As input images with a size of 160 px × 30 px and 3 colour channels are

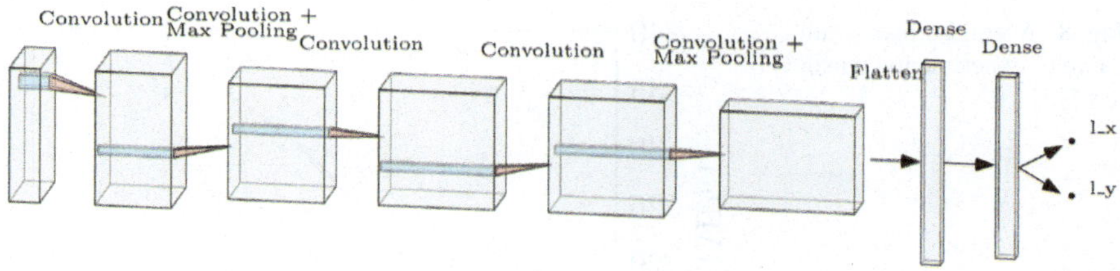

Fig. 7 Schematic depicting the CNN architecture used in this work [73]

Table 1 CNN architecture used in this work

Layer no.	Type and nonlinearity	Input size	Output size
	Input	160 × 30 × 3	160 × 30 × 3
1	Convolution (5 × 5, 16 filters), Batch normalization, Activation (ReLu)	160 × 30 × 3	160 × 30 × 16
2	Convolution (3 × 3, 32 filters), Batch normalization, Activation (ReLu), Max Pooling (2 × 2)	160 × 30 × 16	80 × 15 × 32
3	Convolution (3 × 3, 64 filters), Batch normalization, Activation (ReLu)	80 × 15 × 32	80 × 15 × 64
4	Convolution (3 × 3, 64 filters), Batch normalization, Activation (ReLu)	80 × 15 × 64	80 × 15 × 64
5	Convolution (3 × 3, 128 filters), Batch normalization, Activation (ReLu), Max Pooling (2 × 2)	80 × 15 × 64	40 × 7 × 128
6	Flatten	40 × 7 × 128	35,840
7	Fully connected, Batch normalization, Activation (ReLu), Dropout (0.5)	35,840	2048
8	Fully connected, Batch normalization, Activation (ReLu), Dropout (0.25)	2048	1024
9	2 output layers: Fully connected, Activation (linear)	1024	2 × 1

used. The network has 75.6 million parameters. For implementation of the CNN in Python, the machine learning toolboxes Tensorflow and Keras are used.

3.5 *Experimental*

For predicting the correlation lengths of additively manufactured structures, four tensile tests of additively manufactured specimens are performed utilizing DIC to measure the surface strain distribution. The specimens are manufactured from ABS plastic using Fused Deposition Modeling (FDM) with the specimen geometry accord-

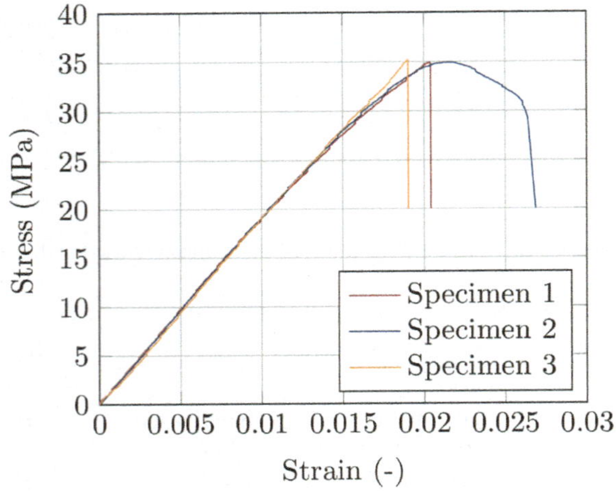

Fig. 8 Average stress-strain curve of experimental tensile test [73]

Table 2 Experimentally determined material parameters

Average Young's modulus	2075 MPa
Young's modulus standard deviation	415 MPa
Average maximum stress	42.1 MPa
Maximum stress standard deviation	2.11 MPa
Poisson's ratio	0.4

ing to ASTM D638 specimen Type IV [9]. The specimen thickness is 1 mm. During printing 100% infill is used with the printing orientation always parallel to the specimen edge. Tests were performed at room temperature and with a displacement rate of 5 mm/min. In Fig. 8 the average stress-strain curve from the performed tensile tests is depicted and Table 2 gives an overview of the determined material parameters.

3.6 Neural Network Training

Using the material parameters given in Table 2, the training dataset for the CNN is created. 100,000 simulations with random correlation lengths ranging from 0.5 to 6 mm are performed. In the training set simulations a strain of 0.5% is applied. The KLE for discretizing the random fields is truncated after 1500 elements. Subsequently, an image of the major strain distribution for each simulation is exported from Paraview using a Python macro. The colour legend for the training images covers two standard deviations around the mean strain value (see Fig. 15) with values higher or lower being the same colour as the maximum or minimum of the legend, respec-

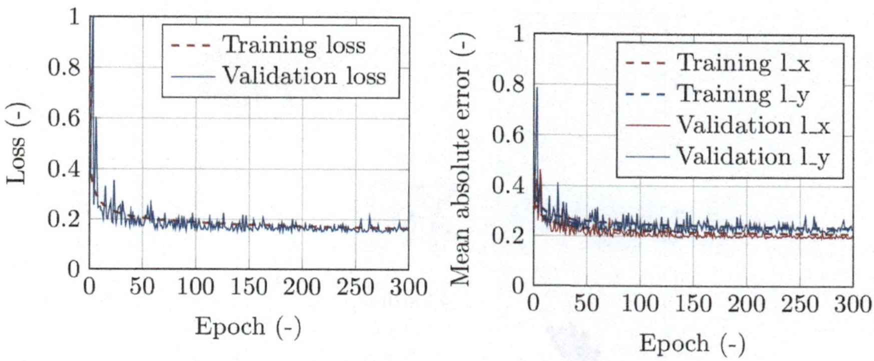

(a) Loss function: Mean squared error.

(b) Mean absolute error.

Fig. 9 Loss function and mean absolute error for both output variables during training [73]

tively. The same colour legend range will later be used for predicting the correlation lengths from DIC images.

The training dataset is split so that 80% of the images are used for training and 20% of the images are used for validation in order to evaluate the training performance. Additionally, data augmentation consisting of horizontal and vertical shifting and horizontal and vertical flipping of the images is used to prevent overfitting [101]. The training data is normalized before training using the Power Transformer from the scikit Python library to improve performance. Training is carried out using the Adam optimiser in Tensorflow [51] with mean squared error as loss function. The model is trained for 300 epochs with a batch size of 50. After finishing 300 epochs, the weights of the best epoch based on validation loss are used.

Figure 9 depicts the loss function progress for training and validation datasets during training and Fig. 9 shows the mean absolute error for both correlation length predictions as additional performance metrics. Improved performance in the predictions of the correlation length in x-direction (i.e. along the width of the specimen) can be observed. After about 100–150 epochs loss and validation loss remain practically stable and the network has reached its generalization capability. An increase in validation loss is not observed, showing that overfitting does not occur.

For validation of the CNN, previously unseen validation images at three different pairs of correlation lengths ($l_1 = l_2 = 1$, $l_1 = l_2 = 3$, $l_1 = l_2 = 5$) with 200 realizations for each length pair are fed through the trained network and the predictions are compared to the expected values. A comparison of the estimated and expected values can be seen in Fig. 10. Over all 600 test images an absolute percentage error of 10.3% is observed. Histograms of the percentage error distributions for both correlation lengths are given in Fig. 11. The best performance is seen at smaller correlation lengths with more scatter at larger correlation lengths.

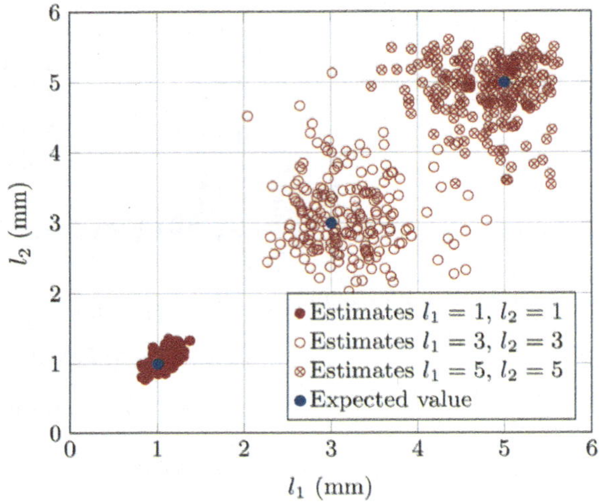

Fig. 10 Comparison of expected and estimated values using a set of validation images [73]

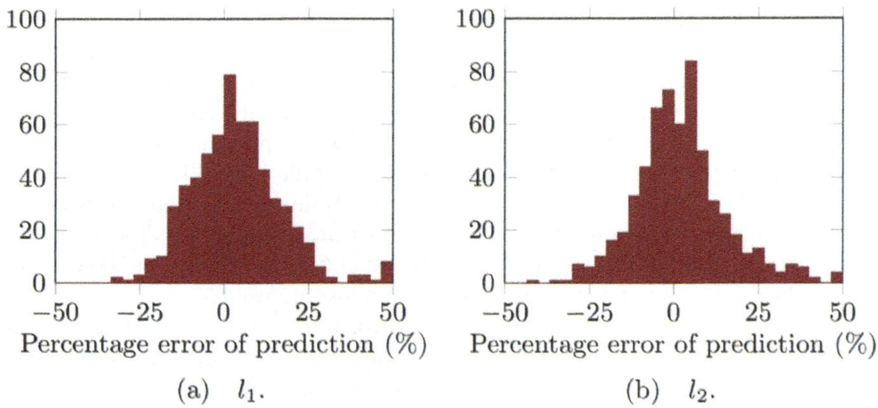

Fig. 11 Histograms of the prediction error for both correlation lengths [73]

It should be noted at this place that the purpose of this research was not to find an optimal NNW architecture for the given problem, but rather show the feasibility of using a NNW for the given problem.

3.7 *Optimization of Neural Network Architecture*

Application of neural networks to engineering problems requires design of suitable network architectures. However, development of high performing neural network architectures commonly requires considerable expertise. Recently, Neural Architecture Search (NAS) methods have shown to produce competitive network architectures with performance comparable or surpassing human developed architectures [29, 58, 72, 99, 106, 107]. To improve performance of the network presented in this chapter, NAS could be used for architecture optimization.

Compared to hyperparameter optimization, which generally requires a good initial model to generate superior network architectures, NAS is more flexible and can design good model architectures from the ground up [106]. Successful developments of NAS methods have been shown in [106, 107] with generated network architectures showing better performance than human designed networks. In [106] a Recurrent Neural Network (RNN) controller is applied to generate network architectures. The controller first generates a sample architecture, which is trained until convergence is reached [72, 106]. Using the performance metrics of the sample architecture, the controller RNN is updated to predict more promising sample architectures. This loop is repeated for multiple iterations. Because NAS requires multiple iterations of fully training different model architectures it is very computationally expensive.

To reduce the computational burden, in [72] Efficient Neural Architecture Search (ENAS) is proposed. Instead of individually training each sample architecture (child model) the authors share weights between all child models. The necessity to fully train the model at each iteration is therefore removed while strong performance is shown. To facilitate shared weights of the child models, the search space is represented as a single Directed Acyclic Graph (DAG) [72]. Nodes in the DAG represent local computations while the edges of the graph represent the flow of information (see Fig. 12) [72]. Therefore, each child model can be represented as a subgraph of the DAG with the parameters shared among all possible architectures.

ENAS uses a Long Short-Term Memory (LSTM) [42] RNN as controller and alternately trains the shared parameters of the child models and the parameters of the controller [72]. During training of the controller parameters, the expected reward function is maximized, which is computed on the validation set of the child models.

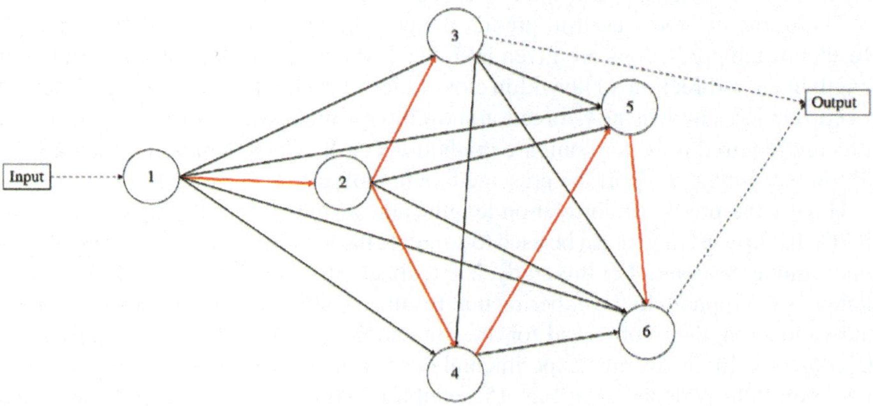

Fig. 12 Representation of the ENAS search space as a DAG with 6 nodes. An arbitrary model in the search space defined by the controller is depicted with red arrows, where node 1 is the input node and the concatenated outputs of nodes 3 and 6 is the model output (adapted from [72])

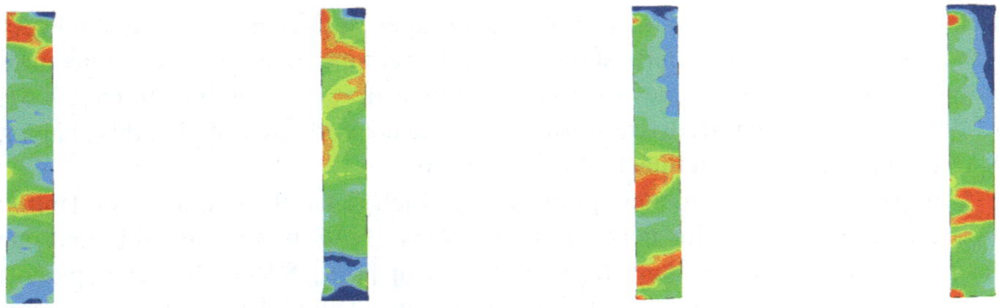

Fig. 13 DIC images used for predicting the experimental correlation lengths and showing major strain distribution during tensile testing [73]

To design a CNN, ENAS allows optimization of individual convolutional and reduction cells and subsequent connection of the individual cells, rather than designing the entire CNN [72]. Nodes in the DAG represent the computations performed in the local cell (identity, convolution with kernel size 3×3 or 5×5, or average or maximum pooling layers). The controller RNN chooses the connections between the nodes and the operations to apply at the sampled nodes [72].

3.8 Estimation of Experimental Correlation Lengths

DIC images are used for estimating the experimentally observed correlation lengths with the trained CNN. As DIC is prone to noise especially at very small strains and only the qualitative strain distribution is of interest for predicting the correlation lengths, DIC images are not evaluated within a fixed strain range but rather within the linear response range once strain localizations visually stabilized. The utilized images from four tensile tests are depicted in Fig. 13.

The trained CNN is used to predict the correlation lengths for all four images. Results of the predictions are given in Table 3. Predicted values for the correlation length in the x-direction all lie within close range of each other, while the predictions for correlation length in the y-direction show more scatter. While further experimental tests are required to better compare predictions of the CNN, similar predictions for all four specimens support the presented methodology.

Having the predicted correlation lengths and with that the fully defined random fields, the forward model can be used to simulate the tensile test and perform statistics on the model response. For this study, 200 realizations of the random fields were simulated and compared to the experimental results (see Fig. 14). Variations in stiffness and yield strength are observed for the simulations as expected and generally good agreement with the average experimental curve can be seen. A single realization of the strain field is depicted in Fig. 15, exhibiting strain localizations similar to the strain field observed using DIC.

Table 3 Predictions for correlation lengths for DIC images

Specimen	l_x	l_y
1	1.75	0.77
2	1.59	1.05
3	1.79	1.40
4	1.68	0.87
Average	1.70	1.02
Standard deviation	0.08	0.28

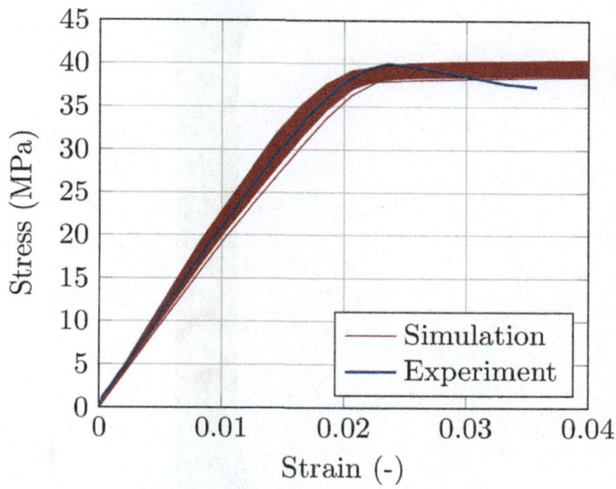

Fig. 14 200 runs of the forward model using the determined parameters and comparison with the experimental curve [73]

4 Concluding Remarks

In this chapter, we adapted a methodology using a neural network to estimate model parameters for modeling stochastic variations of stiffness and strength of additively manufactured structures. KLE was used to create instances of correlated random fields. The covariance function of the two-dimensional random fields requires two correlation lengths. We trained a neural network using strain distributions obtained from simulations with known correlation lengths. Once trained, the network was used to estimate correlation lengths for additively manufactured specimens (solving the inverse problem). With known random field parameters, stochastic simulations with variability of the stiffness and strength can be performed for UQ of printed structures.

Validation of the trained CNNs on previously unseen validation images showed good prediction accuracy for the correlation lengths. However, especially larger correlations lengths displayed relatively large prediction errors. Prediction accuracy for large correlation lengths could be improved in future work by using a larger simulation domain to better capture the effects of large correlation lengths. Finally, the performance of the networks could be improved using NAS.

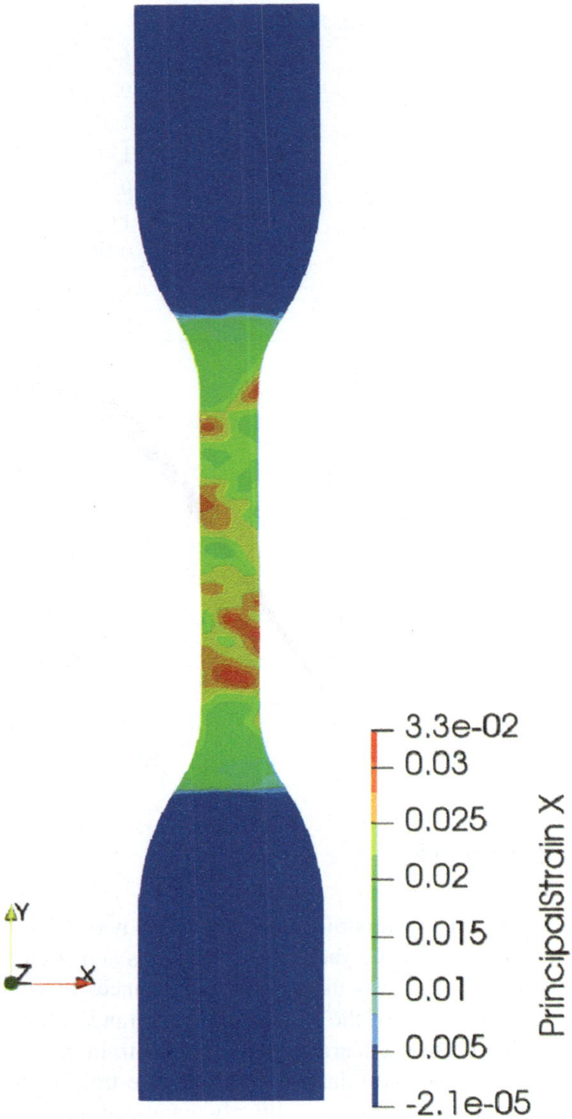

Fig. 15 Major strain distribution for a realization of the random field using the predicted correlation lengths at an applied strain of 2% [73]

Future work will include performance improvements of the presented methodology. Additional experimental effort will also be necessary to validate the estimated parameters.

References

1. Abbassi F, Belhadj T, Mistou S, Zghal A (2013) Parameter identification of a mechanical ductile damage using artificial neural networks in sheet metal forming. Mater Des 45:605–615. ISSN: 0261-3069. https://doi.org/10.1016/j.matdes.2012.09.032
2. Abiodun OI, Jantan A, Omolara AE, Dada KV, Umar AM, Linus OU, Arshad H, Kazaure AA, Gana U, Kiru MU (2019) Comprehensive review of artificial neural network applications to pattern recognition. IEEE Access 7:158820–158846
3. Adams BM, Bohnhoff WJ, Dalbey KR, Ebeida MS, Eddy JP, Eldred MS, Hooper RW et al (2021) Dakota, a multilevel parallel object-oriented framework for design optimization, parameter estimation, uncertainty quantification, and sensitivity analysis: version 6.14 user's manual. Sandia National Laboratories
4. Adler RJ, Taylor JE (2007) Random fields and geometry. Springer monographs in mathematics. Springer, New York, p 115. ISBN: 9780387481128
5. Alazwari Mashhour A, Rao Singiresu S (2018) Interval-based uncertainty models for micromechanical properties of composite materials. J Reinf Plast Compos 37(18):1142–1162. https://doi.org/10.1177/0731684418788733
6. Anders M, Hori M (2001) Three-dimensional stochastic finite element method for elasto-plastic bodies. Int J Numer Methods Eng 51(4):449–478. https://doi.org/10.1002/nme.165
7. Apra M, D'Amore M, Gigliotti K, Maria SS, Volpi V (2008) Lightning indirect effects certification of a transport aircraft by numerical simulation. IEEE Trans Electromagn Compat 50(3):513–523. https://doi.org/10.1109/TEMC.2008.927738
8. Arregui-Mena José D, Margetts L, Mummery Paul M (2016) Practical application of the stochastic finite element method. Arch Comput Methods Eng 23(1):171–190. https://doi.org/10.1007/s11831-014-9139-3
9. ASTM International (2014) ASTM D638-14, Standard test method for tensile properties of plastics. ASTM International, West Conshohocken, PA. https://doi.org/10.1520/D0638-14. http://www.astm.org
10. Balokas G, Kriegesmann B, Czichon S, Rolfes R (2021) A variable-fidelity hybrid surrogate approach for quantifying uncertainties in the nonlinear response of braided composites. Comput Methods Appl Mech Eng 381:113851. ISSN: 0045-7825. https://doi.org/10.1016/j.cma.2021.113851
11. Balokas G, Kriegesmann B, Rolfes R (2021) Data-driven inverse uncertainty quantification in the transverse tensile response of carbon fiber reinforced composites. Compos Sci Technol 211:108845. ISSN: 0266-3538. https://doi.org/10.1016/j.compscitech.2021.108845
12. Bengio Y (2009) Learning deep architectures for AI. Found Trends Mach Learn 2(1):1–127. ISSN: 1935-8237. https://doi.org/10.1561/2200000006
13. Borja RI (2013) J2 plasticity. In: Plasticity: modeling and computation. Springer, Berlin, pp 31–58. ISBN: 978-3-642-38547-6. https://doi.org/10.1007/978-3-642-38547-6
14. Bostanabad R, Bui AT, Xie W, Apley DW, Chen W (2016) Stochastic microstructure characterization and reconstruction via supervised learning. Acta Mater 103:89–102. ISSN: 1359-6454. https://doi.org/10.1016/j.actamat.2015.09.044
15. Bostanabad R, Liang B, Gao J, Liu WK, Cao J, Zeng D, Su X, Xu H, Li Y, Chen W (2018) Uncertainty quantification in multiscale simulation of woven fiber composites. Comput Methods Appl Mech Eng 338:506–532. ISSN: 0045-7825. https://doi.org/10.1016/j.cma.2018.04.024

16. Bratley P, Fox BL (1988) Algorithm 659: implementing Sobol's quasirandom sequence generator. ACM Trans Math Softw 14(1):88–100. ISSN: 0098-3500. https://doi.org/10.1145/42288.214372
17. Burkardt J (2019) SOBOL FORTRAN90 library. https://github.com/johannesgerer/jburkardt-f/tree/master/sobol
18. Cayot JE, Harper CW (1976) Simulation's potential role in advanced aircraft certification. SAE Tech Pap. https://doi.org/10.4271/760931
19. Chamekh A, Bel Hadj Salah H, Hambli R (2008) Inverse technique identification of material parameters using finite element and neural network computation. Int J Adv Manuf Technol 44(1):173. https://doi.org/10.1007/s00170-008-1809-6
20. Chandrupatla TR, Belegundu AD (2012) Introduction to finite elements in engineering, 4th edn. Pearson. ISBN: 9780132162746
21. Charmpis DC, Schuëller GI, Pellissetti MF (2007) The need for linking micromechanics of materials with stochastic finite elements: a challenge for materials science. Comput Mater Sci 41(1):27–37. ISSN: 0927-0256. https://doi.org/10.1016/j.commatsci.2007.02.014
22. Chu S, Guilleminot J, Kelly C, Abar B, Gall K (2021) Stochastic modeling and identification of material parameters on structures produced by additive manufacturing. Comput Methods Appl Mech Eng 387:114166. ISSN: 0045-7825. https://doi.org/10.1016/j.cma.2021.114166
23. Cizelj L, Mavko B, Riesch-Oppermann H (1994) Application of first and second order reliability methods in the safety assessment of cracked steam generator tubing. Nucl Eng Des 147(3):359–368
24. Cizelj L, Riesch-Oppermann H (1992) ZERBERUS the code for reliability analysis of crack containing structures
25. Cui C, Fearn T (2018) Modern practical convolutional neural networks for multivariate regression: applications to NIR calibration. Chemom Intell Lab Syst 182:9–20. ISSN: 0169-7439. https://doi.org/10.1016/j.chemolab.2018.07.008
26. Dang HV, Trestian R, Bui-Tien T, Nguyen HX (2021) Probabilistic method for time-varying reliability analysis of structure via variational Bayesian neural network. Structures 34:3703–3715. ISSN: 2352-0124. https://doi.org/10.1016/j.istruc.2021.09.069
27. Der Kiureghian A, Haukaas T, Fujimura K (2006) Structural reliability software at the University of California, Berkeley. Struct Saf 28(1):44–67. ISSN: 0167-4730. https://doi.org/10.1016/j.strusafe.2005.03.002
28. Dey S, Mukhopadhyay T, Adhikari S (2019) Uncertainty quantification in laminated composites: a meta-model approach. CRC Press, Taylor and Francis Group. ISBN: 9781498784450
29. Elsken T, Metzen JH, Hutter F (2019) Neural architecture search: a survey. arXiv: 1808.05377
30. Fischer P, Dosovitskiy A, Brox T (2015) Image orientation estimation with convolutional networks. In: German conference on pattern recognition (GCPR). Springer, Berlin
31. Fu B, Yang Y, Qi H, Xu J, Wang S (2021) Modeling and analysis of high aspect ratio wing considering random structural parameters. Sci Rep 11(1):15609. https://doi.org/10.1038/s41598-021-95187-0
32. Fuglstad G-A, Simpson D, Lindgren F, Rue H (2015) Does non-stationary spatial data always require non-stationary random fields? Spat Stat 14:505–531. ISSN: 2211-6753. https://doi.org/10.1016/j.spasta.2015.10.001
33. Fukunaga K, Koontz WLG (1970) Representation of random processes using the finite Karhunen-Loève expansion. Inf Control 16(1):85–101. ISSN: 0019-9958. https://doi.org/10.1016/S0019-9958(70)80043-2
34. Gardiner G (2014) The path toward certification by simulation. CompositesWorld. https://www.compositesworld.com/articles/the-path-towardcertification-by-simulation-part-2-uq-lessons-from-nnsa
35. Georgiadis S, Gunnion AJ, Thomson RS, Cartwright BK (2008) Bird-strike simulation for certification of the Boeing 787 composite moveable trailing edge. In: Fourteenth international conference on composite structures. Compos Struct 86(1):258–268. ISSN: 0263-8223. https://doi.org/10.1016/j.compstruct.2008.03.025

36. Ghanem Roger G, Pol S (1991) Stochastic finite elements: a spectral approach. Springer, New York. ISBN 978-1-4612-7795-8
37. Glasserman P (2004) Monte Carlo methods in financial engineering. Appl Math 53. ISBN: 9780387004518
38. Gupta A, Cecen A, Goyal S, Singh AK, Kalidindi SR (2015) Structure-property linkages using a data science approach: application to a non-metallic inclusion/steel composite system. Acta Mater 91:239–254. ISSN: 1359-6454. https://doi.org/10.1016/j.actamat.2015.02.045
39. Gupta KK, Mukhopadhyay T, Roy A, Roy L, Dey S (2021) Sparse machine learning assisted deep computational insights on the mechanical properties of graphene with intrinsic defects and doping. J Phys Chem Solids 155:110111. ISSN: 0022-3697. https://doi.org/10.1016/j.jpcs.2021.110111
40. Haldar A, Mahadevan S (2000) Probability, reliability, and statistical methods in engineering design. Wiley, Hoboken
41. Hesterberg TC (2015) What teachers should know about the bootstrap: resampling in the undergraduate statistics curriculum. Am Stat 69(4):371–386. https://doi.org/10.1080/00031305.2015.1089789. PMID: 27019512
42. Hochreiter S, Schmidhuber J (1997) Long short-term memory. Neural Comput 9(8):1735–1780. ISSN: 0899-7667. https://doi.org/10.1162/neco.1997.9.8.1735
43. Huang T, Gao J, Sun Q, Zeng D, Su X, Liu WK, Chen W (2021) Stochastic nonlinear analysis of unidirectional fiber composites using image-based microstructural uncertainty quantification. Compos Struct 260:113470. ISSN: 0263-8223. https://doi.org/10.1016/j.compstruct.2020.113470
44. Impollonia N, Sofi A (2003) A response surface approach for the static analysis of stochastic structures with geometrical nonlinearities. Comput Methods Appl Mech Eng 192(37):4109–4129. ISSN: 0045-7825. https://doi.org/10.1016/S0045-7825(03)00379-7
45. Intel (2021) Developer reference for Intel one API math kernel library—Fortran. https://www.intel.com/content/www/us/en/develop/documentation/onemkldeveloper-reference-fortran/top.html
46. Ioffe S, Szegedy C (2015) Batch normalization: accelerating deep network training by reducing internal covariate shift. arXiv: 1502.03167
47. Jäckel P (2002) Monte Carlo methods in finance. The Wiley finance series. Wiley, New York. ISBN: 978-0-471-49741-7
48. Jeong HK, Shenoi RA (2000) Probabilistic strength analysis of rectangular FRP plates using Monte Carlo simulation. Comput Struct 76(1):219–235. ISSN: 0045-7949. https://doi.org/10.1016/S0045-7949(99)00171-6
49. Kalagnanam Jayant R, Diwekar Urmila M (1997) An efficient sampling technique for off-line quality control. Technometrics 39(3):308–319. https://doi.org/10.1080/00401706.1997.10485122
50. Kasinos S, Palmeri A, Lombardo M, Adhikari S (2021) A reduced modal subspace approach for damped stochastic dynamic systems. Comput Struct 257:106651. ISSN: 0045-7949. https://doi.org/10.1016/j.compstruc.2021.106651
51. Kingma DP, Ba J (2015) Adam: a method for stochastic optimization. In: Bengio Y, LeCun Y (eds) 3rd international conference on learning representations, ICLR 2015, San Diego, CA, USA, May 7–9, 2015, Conference track proceedings
52. Klenke A (2014) Probability theory: a comprehensive course. Springer, London. ISBN: 978-1-4471-5361-0. https://doi.org/10.1007/978-1-4471-5361-0
53. Krizhevsky A, Sutskever I, Hinton GE (2017) ImageNet classification with deep convolutional neural networks. Commun ACM 60(6):84–90. ISSN: 0001-0782. https://doi.org/10.1145/3065386
54. Kucherenko S, Albrecht D, Saltelli A (2015) Exploring multidimensional spaces: a comparison of Latin hypercube and quasi Monte Carlo sampling techniques. arXiv: 1505.02350 [stat.AP]
55. Lähivaara T, Kärkkäinen L, Huttunen Janne M, Hesthaven Jan S (2018) Deep convolutional neural networks for estimating porous material parameters with ultrasound tomography. J Acoust Soc Am 143(2):1148–1158. https://doi.org/10.1121/1.5024341

56. LeCun Y, Bengio Y, Hinton G (2015) Deep learning. Nature 521(7553):436–444. ISSN: 1476-4687. https://doi.org/10.1038/nature14539
57. Lekou DJ, Philippidis TP (2008) Mechanical property variability in FRP laminates and its effect on failure prediction. Compos Part B Eng 39(7):1247–1256. ISSN: 1359-8368. https://doi.org/10.1016/j.compositesb.2008.01.004
58. Liu H, Simonyan K, Yang Y (2018) DARTS: differentiable architecture search. arXiv: 1806.09055
59. Lykkegaard MB, Dodwell TJ, Moxey D (2021) Accelerating uncertainty quantification of groundwater flow modelling using a deep neural network proxy. Comput Methods Appl Mech Eng 383:113895. ISSN: 0045-7825. https://doi.org/10.1016/j.cma.2021.113895
60. Mahendran S, Ali H, Vidal R (2017) 3D PoseRegression using convolutional neural networks. In: 2017 IEEE conference on computer vision and pattern recognition workshops (CVPRW), pp 494–495
61. Meißner P, Watschke H, Winter J, Vietor T (2020) Artificial neural networks-based material parameter identification for numerical simulations of additively manufactured parts by material extrusion. Polymers 12(12). ISSN: 2073-4360. https://doi.org/10.3390/polym12122949
62. Mohamedou M, Zulueta K, Chung CN, Rappel H, Beex L, Adam L, Arriaga A, Major Z, Wu L, Noels L (2019) Bayesian identification of mean-field homogenization model parameters and uncertain matrix behavior in non-aligned short fiber composites. Compos Struct 220:64–80. ISSN: 0263-8223. https://doi.org/10.1016/j.compstruct.2019.03.066
63. Montoya-Noguera S, Zhao T, Hu Y, Wang Y, Phoon K-K (2019) Simulation of non-stationary non-Gaussian random fields from sparse measurements using Bayesian compressive sampling and Karhunen-Loève expansion. Structu Safe 79:66–79. ISSN: 0167-4730. https://doi.org/10.1016/j.strusafe.2019.03.006
64. Nader Jacques W, Dagher Habib J, El Chiti F, Lopez-Anido R (2009) Probabilistic finite element analysis of ASTM D6641 compression test for marine grade polymer matrix composites. J Reinf Plast Compos 28(8):897–911. https://doi.org/10.1177/0731684407087372
65. Nader Jacques W, Dagher Habib J, Lopez-Anido R, El Chiti F, Fayad GN, Thomson L (2008) Probabilistic finite element analysis of modified ASTM D3039 tension test for marine grade polymer matrix composites. J Reinf Plast Compos 27(6):583–597. https://doi.org/10.1177/07316844070799152008
66. Nassef MGA, Schenck C, Kuhfuss B (2011) Simulation-based parameter identification of a reduced model using neural networks. In: 2011 9th IEEE international conference on control and automation (ICCA), pp 974–978. https://doi.org/10.1109/ICCA.2011.6138002
67. National Research Council (U.S.), Committee on Mathematical Foundations of Verification, Validation, and Uncertainty Quantification (2012) Assessing the reliability of complex models: mathematical and statistical foundations of verification, validation, and uncertainty quantification. National Academies Press. ISBN: 9780309256353
68. Niederreiter H (1988) Low-discrepancy and low-dispersion sequences. J Number Theor 30(1):51–70. ISSN: 0022-314X. https://doi.org/10.1016/0022-314X(88)90025-X
69. Oden JT, Belytschko T, Babuska I, Hughes TJR (2003) Research directions in computational mechanics. Comput Methods Appl Mech Eng 192(7):913–922. ISSN: 0045-7825. https://doi.org/10.1016/S0045-7825(02)00616-3
70. Olfatbakhsh T, Milani AS (2022) A highly interpretable materials informatics approach for predicting microstructure-property relationship in fabric composites. Compos Sci Technol 217:109080. ISSN: 0266-3538. https://doi.org/10.1016/j.compscitech.2021.109080. https://www.sciencedirect.com/science/article/pii/S026635382100436X
71. Olivier A, Shields MD, Graham-Brady L (2021) Bayesian neural networks for uncertainty quantification in data-driven materials modeling. Comput Methods Appl Mech Eng 386:114079. ISSN: 0045-7825. https://doi.org/10.1016/j.cma.2021.114079
72. Pham H, Guan MY, Zoph B, Le QV, Dean J (2018) Efficient neural architecture search via parameter sharing. arXiv: 1802.03268
73. Pitz EJ, Rooney SE, Pochiraju KV (2020) Stochastic modelling of additively manufactured structures using a neural network for identification of random field parameters. In: Proceedings of the 35th ASC conference. https://doi.org/10.12783/asc35/34974

74. Rice JA (1944/1995) Mathematical statistics and data analysis, 2nd edn. Duxbury advanced series. Duxbury Press, Belmont. ISBN: 9780534209346
75. Singhee A, Rutenbar RA (2007) From finance to flip flops: a study of fast quasi-Monte Carlo methods from computational finance applied to statistical circuit analysis. In: 8th international symposium on quality electronic design (ISQED'07), pp 685–692. https://doi.org/10.1109/ISQED.2007.79
76. Singhee A, Rutenbar Rob A (2010) Why quasi-Monte Carlo is better than Monte Carlo or Latin hypercube sampling for statistical circuit analysis. IEEE Trans Comput-Aided Des Integr Circ Syst 29(11):1763–1776. https://doi.org/10.1109/TCAD.2010.2062750
77. Sobol IM (1967) On the distribution of points in a cube and the approximate evaluation of integrals. USSR Comput Math Math Phys 7(4):86–112. ISSN: 0041-5553. https://doi.org/10.1016/0041-5553(67)90144-9
78. Sriramula S, Chryssanthopoulos MK (2009) Quantification of uncertainty modelling in stochastic analysis of FRP composites. Compos Part A Appl Sci Manuf 40(11):1673–1684. ISSN: 1359-835X. https://doi.org/10.1016/j.compositesa.2009.08.020
79. Srivastava N, Hinton G, Krizhevsky A, Sutskever I, Salakhutdinov R (2014) Dropout: a simple way to prevent neural networks from overfitting. J Mach Learn Res 15(1):1929–1958. ISSN: 1532-4435
80. Staber B, Guilleminot J, Soize C, Michopoulos J, Iliopoulos A (2019) Stochastic modeling and identification of a hyperelastic constitutive model for laminated composites. Comput Methods Appl Mech Eng 347:425–444. ISSN: 0045-7825. https://doi.org/10.1016/j.cma.2018.12.036
81. Stefanou G (2009) The stochastic finite element method: past, present and future. Comput Methods Appl Mech Eng 198(9):1031–1051. ISSN: 0045-7825. Accessed 3 Feb 2019. https://doi.org/10.1016/j.cma.2008.11.007
82. Sullivan TJ (2015) Introduction to uncertainty quantification: texts in applied mathematics, vol 63. Springer. ISBN: 9783319233956. https://doi.org/10.1007/978-3-319-23395-6
83. Tal D, Fish J (2016) Generating a statistically equivalent representative volume element with discrete defects. Compos Struct 153:791–803. ISSN: 0263-8223. https://doi.org/10.1016/j.compstruct.2016.06.077
84. Tao W, Zhu P, Xu C, Liu Z (2020) Uncertainty quantification of mechanical properties for three-dimensional orthogonal woven composites. Part I: stochastic reinforcement geometry reconstruction. Compos Struct 235:111763. ISSN: 0263-8223. https://doi.org/10.1016/j.compstruct.2019.111763
85. Tao W, Zhu P, Xu C, Liu Z (2020) Uncertainty quantification of mechanical properties for threedimensional orthogonal woven composites. Part II: multiscale simulation. Compos Struct 235:111764. ISSN: 0263-8223. https://doi.org/10.1016/j.compstruct.2019.111764
86. Vanaerschot A, Cox BN, Lomov SV, Vandepitte D (2013) Stochastic multi-scale modelling of textile composites based on internal geometry variability. Computational fluid and solid mechanics. Comput Struct 122:55–64. ISSN: 0045-7949. https://doi.org/10.1016/j.compstruc.2012.10.026
87. Verderaime V, Marshall George C, Flight Center Space (1994) Illustrated structural application of universal first-order reliability method. NASA technical paper, National Aeronautics/Space Administration, Marshall Space Flight Center
88. Vittaldev V, Russell Ryan P, Linares R (2016) Spacecraft uncertainty propagation using Gaussian mixture models and polynomial chaos expansions. J Guidance Control Dyn 39(12):2615–2626. https://doi.org/10.2514/1.G001571
89. Vořechovský M (2008) Simulation of simply cross correlated random fields by series expansion methods. Struct Saf 30(4):337–363. ISSN: 0167-4730. https://doi.org/10.1016/j.strusafe.2007.05.002
90. Wang R, Diwekar U, Grégoire Padró CE (2004) Efficient sampling techniques for uncertainties in risk analysis. Environ Prog 23(2):141–157. https://doi.org/10.1002/ep.10020
91. Wang X, Sloan IH (2008) Low discrepancy sequences in high dimensions: how well are their projections distributed? J Comput Appl Math 213(2):366–386. ISSN: 0377-0427. https://doi.org/10.1016/j.cam.2007.01.005

92. Wiener N (1938) The homogeneous chaos. Am J Math 60(4):897–936. ISSN: 00029327, 10806377. http://www.jstor.org/stable/2371268
93. Wu L, Adam L, Noels L (2018) A micromechanics-based inverse study for stochastic order reduction of elastic UD fiber reinforced composites analyses. Int J Numer Methods Eng 115(12):1430–1456. ISSN: 1097-0207. https://doi.org/10.1002/nme.5903
94. Wu L, Chung CN, Major Z, Adam L, Noels L (2018) From SEM images to elastic responses: a stochastic multiscale analysis of UD fiber reinforced composites. Compos Struct 189:206–227. ISSN: 0263-8223. https://doi.org/10.1016/j.compstruct.2018.01.051
95. Wu L, Nguyen V-D, Adam L, Noels L (2019) An inverse micro-mechanical analysis toward the stochastic homogenization of nonlinear random composites. Comput Methods Appl Mech Eng 348:97–138. ISSN: 0045-7825. https://doi.org/10.1016/j.cma.2019.01.016
96. Wu L, Nguyen VD, Kilingar NG, Noels L (2020) A recurrent neural network-accelerated multi-scale model for elasto-plastic heterogeneous materials subjected to random cyclic and non-proportional loading paths. Comput Methods Appl Mech Eng 369:113234. ISSN: 0045-7825. https://doi.org/10.1016/j.cma.2020.113234
97. Wu W-F, Cheng H-C, Kang C-K (2000) Random field formulation of composite laminates. Compos Struct 49(1):87–93. ISSN: 0263-8223. https://doi.org/10.1016/S0263-8223(99)00128-2
98. Xiao H, Cinnella P (2019) Quantification of model uncertainty in RANS simulations: a review. Prog Aerosp Sci 108:1–31. ISSN: 0376-0421. https://doi.org/10.1016/j.paerosci.2018.10.001
99. Xie L, Chen X, Bi K, Wei L, Xu Y, Wang L, Chen Z et al (2021) Weight-sharing neural architecture search: a battle to shrink the optimization gap. ACM Comput Surv 54(9). ISSN: 0360-0300. https://doi.org/10.1145/3473330
100. Xun L, Yuwen J, Tong Z, Liang Z (2017) Credit taken for simulation system used in aircraft type certification. In: 2017 8th international conference on mechanical and aerospace engineering (ICMAE), pp 517–521. https://doi.org/10.1109/ICMAE.2017.8038700
101. Yamashita R, Nishio M, Do RKG, Togashi K (2018) Convolutional neural networks: an overview and application in radiology. Insights Imaging 9(4):611–629. ISSN: 1869-4101. https://doi.org/10.1007/s13244-018-0639-9
102. Young GA, Smith RL (2005) Bootstrap methods. In: Essentials of statistical inference. Cambridge series in statistical and probabilistic mathematics. Cambridge University Press, pp 190–217. https://doi.org/10.1017/CBO9780511755392.012
103. Yun GJ, Zhao L, Iarve E (2015) Probabilistic mesh-independent discrete damage analyses of laminate composites. Compos Struct 133:22–30. ISSN: 0263-8223. https://doi.org/10.1016/j.compstruct.2015.07.083
104. Zhang M, Liu X, Wang Y, Wang X (2019) Parameter distribution characteristics of material fatigue life using improved bootstrap method. Int J Damage Mech 28(5):772–793. https://doi.org/10.1177/1056789518792658
105. Zhao T, Wang Y (2018) Simulation of cross-correlated random field samples from sparse measurements using Bayesian compressive sensing. Mech Syst Sig Process 112:384–400. ISSN: 0888-3270. https://doi.org/10.1016/j.ymssp.2018.04.042
106. Zoph B, Le QV (2016) Neural architecture search with reinforcement learning. arXiv: 1611.01578
107. Zoph B, Vasudevan V, Shlens J, Le QV (2017) Learning transferable architectures for scalable image recognition. arXiv: 1707.07012

Radial Basis Function-Based Uncertain Low-Velocity Impact Behavior Analysis of Functionally Graded Plates

P. K. Karsh, R. R. Kumar, Vaishali, and S. Dey

1 Introduction

Functionally graded material changes its material properties, density, composition, and microstructure across the volume. It is a hybrid of composite material in which two dissimilar materials, namely ceramic and metal, are non-homogenously intermixed to form a graded structure [1]. They are designed to perform specified functions with automobiles, biomedical, marine, aerospace, and civil applications. In bio-medical applications, implants are made by FGM because most parts of the human body are made of FGM-like structures. The conventional composite materials are produced by homogenous elastic laminates that bond together to intensify the mechanical properties. The material properties at the interface of two laminates are abruptly changed. This can create enormous intermediate stress, leading to the delamination of laminates. The phenomena of delamination mainly occur at the higher load and temperature during the service of the composite structure. This lacuna of the composite structure is overcome by using FGMs in which there is no interface inside the system. So, the stress distribution inside the system is smooth. The FGM structures are mainly employed where one side of the plate requires sufficient thermal resistance while the second side of the plate requires good yield strength [2]. They are graded structures in which two materials are distributed following

P. K. Karsh (✉)
Department of Mechanical Engineering, Parul Institute of Engineering and Technology, Parul University, Vadodara, India
e-mail: pradeepkarsh@gmail.com

R. R. Kumar
Department of Aeronautical Engineering, Hindustan Institute of Technology and Science, Chennai, India

Vaishali · S. Dey
Department of Mechanical Engineering, National Institute of Technology Silchar, Silchar, India
e-mail: sudip@mech.nits.ac.in

V. Kushvaha et al. (eds.), *Machine Learning Applied to Composite Materials*,
Composites Science and Technology, https://doi.org/10.1007/978-981-19-6278-3_4

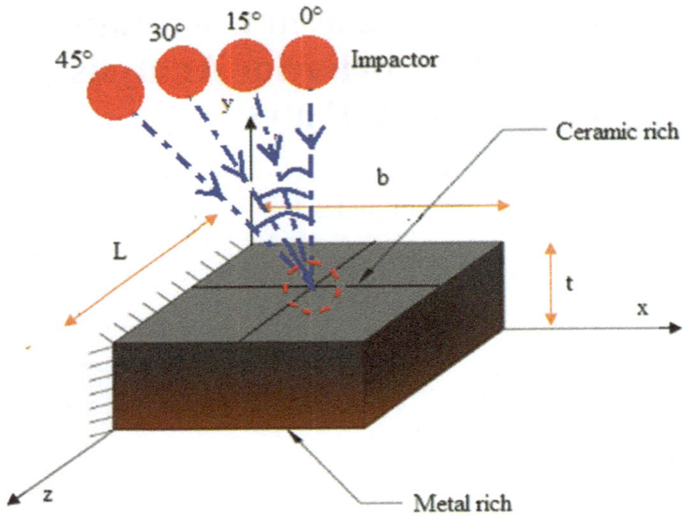

Fig. 1 FGM plate subjected to impact loading

various laws such as power law, sigmoid law, and exponential law. The present model consists of a cantilever plate structure with top surface ceramic rich. In contrast, the bottom surface is metal-rich, and the various other material properties are distributed following power-law (refer to Fig. 1). The distribution pattern of materials inside the plate can be controlled by the controlling parameter called exponent index (power-law index in case of power-law). FGM's material properties depend on the spatial position of the individual material on the bulk material so that FGM has varying material properties on the entire volume.

Many FGM structure applications are subjected to impact loading during the service life, starting from manufacturing to end-use. Due to this impact loading, deformation may occur in the plate and impactor, affecting the FGM structure's functionality. So, it is necessary to analyze the characteristic of FGM structure due to impact loading. In the past, composites and FGM have obtained considerable attention in research work in impact problems, such as Soto et al. [3] worked on low-velocity impact and compression behavior of composite by employing the numerical approach to find the damage resistance of the panel. Hsieh et al. [4] conducted experiments on nanocarbon aerogels for determining the impact and non-impact properties, considering different impact energies. Thorsson et al. [5] employed a shell-based finite element method to determine the reinforced composite's impact behavior and failure mechanism. He also defined fiber rapture and matrix cracking by using the non-destructive inspection method. Schwab et al. [6] analyzed the high-velocity impact response of composite laminated panels, considering different impact velocities and impact angles. The results illustrate that the results obtained from experiments and simulations have fair agreements. Sun et al. [7] conducted a comparison study between thermoset and thermoplastic polymer composites' impact responses by employing numerical and experimental approaches. Elamin et al. [8] worked on the impact of a composite sandwich structure subjected to the very low-temperature environment known as an arctic condition. The experimental results illustrate that the extremely low temperature significantly affects the damage behavior of the sandwich composites. Similarly, Karsh et al. [9] worked on FGM plates and carried out probabilistic and non-probabilistic impact responses by employing the artificial neural network and fuzzy approach. Mao et al. [10] determined the FGM coated shallow spherical shell's interfacial damage behavior experiencing low-velocity impact loading by employing the numerical method. Kiani et al. [11] worked on the FG-beam structure and analyzed the low-velocity impact characteristic using the numerical approach considering the thermal effects. Runge–Kutta approach and Hertz contact law are applied to find the impact of different parameters on the FGM beam's impact responses. Najafi et al. [12] utilized HOSDT for determining the nonlinear impact behavior of the FGM beam subjected to noncentral loading. The modified Hertz's contract law (HCL) is applied to assess the impact behaviors. Eghtesad et al. [13] worked on FGM considering sigmoid law for material properties gradation and analyzed high-velocity impact behavior. Free vibration, bending, and buckling responses for graded material [14–16] and low-velocity impact [17–19] responses for several combinations of composites have been studied using a deterministic approach. Dey et al. [20, 21] analyzed low-velocity impact behavior by employing FE methods on conical shells considering delamination effects. Similarly, in the past, the radial basis function model was applied in many engineering fields for uncertainty analysis, such as Liu et al. [22] for analyzing the static behavior considering uncertainty in input parameters. Er et al. [23] carried out machine vibration analysis using the modal analysis and the radial basis function approach. Li et al. [24] also applied RBF for the structure's reliability analysis, while [25] used it to analyze global sensitivity. Some researches are also carried out considering stochastic nature in various structures [9, 26, 27].

In the last paragraph, a thorough literature review is presented in which most of the research was carried out in deterministic regimes. As discussed earlier, FGM has a nonhomogeneous mixture of two materials; therefore, it can be considered as a highly complex structure. From the design stage to the manufacturing stage of FGM, there are many uncertainties involved, which results in the deviation in values from their deterministic value. These deviations may affect the impact responses of the FGM structure. FGM structure's impact responses depend upon the material and geometrical properties of the plate and impact parameters such as impact angle. Here, the FGM plate's stochastic transient low-velocity impact parameters are determined by considering the variabilities in material properties by employing a surrogate-based RBF model. As already known, MCS is the conventional method to handle probabilistic as well as non-probabilistic uncertainty quantification problems. The MCS considered every possible combination of input variables leads to low computational efficiency. The traditional MCS required thousands of finite element simulations, which leads to high cost and time. In a layered/graded structure such as FGM, finite element simulation is an expansive process that requires time and cost. To mollify with this demerit of the conventional MCS, a surrogate-based RBF is introduced in this chapter. By doing this, the number of FE simulations required is significantly less than traditional MCS, leading to higher computational performance without affecting the meticulousness of the results.

2 Mathematical Formulation

The equation for dynamic analysis is given by Dey et al. [28]

$$[M(\hat{\omega})]\{\ddot{\delta}\} + [K(\hat{\omega})]\{\delta\} = \{F\} \tag{1}$$

The contact force $\{F_C^*\}$ in the impact loading can be represented as

$$\{F\} = \{0\ 0\ 0\ \ldots\ F_C^*\ \ldots\ 0\ 0\ 0\}^T \tag{2}$$

The equation the impactor motion is given in Eq. (3)

$$m_{(I)}\ddot{\delta}_{(I)} + F_C^* = 0 \tag{3}$$

wherein $m_{(I)}$ and $\ddot{\delta}_{(I)}$ denotes impactor's mass and acceleration, respectively. The force at the time when the elastic impactor touches the FGM plate is called the contact force and is determined by employing the modified Hertzian contact (MHC). Here, the impact loading is applied at the middle of the plate, as shown in Fig. 1. The contact force $\left(F_C^*\right)$ for that is given by Sun and Chen [29].

$$F_C^* = K_M \psi^{1.5} \quad 0 < \psi \leq \psi_{\max} \tag{4}$$

$$K_M = \frac{16}{3\pi} \frac{1}{K_P + K_I} \sqrt{\frac{B}{\lambda}} \tag{5}$$

where K_M, ψ, and $\psi_{\max}$ represent modified contact stiffness, local indentation, and maximum indentation, respectively [30]. C and λ are the constant, and for the present study, $\lambda = 2$ is considered. The local indentation can be obtained by

$$\psi(t) = \delta_I(t)\mathrm{Cos}\theta - \delta_P(x_c, y_c, t)\mathrm{Cos}\phi \tag{6}$$

where δ_I is impactor displacement and δ_P is plate displacements, at point of loading (x_c, y_c) in the z-direction, while θ is oblique impact angle and ϕ is twist angle. At the loading point, the contact force components are

$$F_{ix} = 0, \quad F_{iy} = 0, \quad F_{iz} = F_C^* \tag{7}$$

The transient nature of the contact force (F_C^*) can be observed with Δt time intervals. Here, Newmark's integration scheme [31] is employed for evaluating time-dependent equations. By applying the Newmark's integration scheme in Eqs. (1) and (3), succeeding equations at $(t + \Delta t)$ time can be obtained as

$$[\tilde{K}_P]\{\delta\}^{t+\Delta t} = \left\{\tilde{F}_C^*\right\}^{t+\Delta t} \tag{8}$$

$$[\tilde{K}_I]\{\delta_I\}^{t+\Delta t} = \left\{\tilde{F}_C^*\right\}^{t+\Delta t} \tag{9}$$

where effective stiffness matrix for target plate is $[\tilde{K}_P]$ and for impactor is $[\tilde{K}_I]$ and is represented as

$$\left[\tilde{K}_P\right] = [K(\overline{\omega})] + [K_\sigma] + c_0[M(\overline{\omega})] \tag{10}$$

$$\left[\tilde{K}_I\right] = c_0 m_I \tag{11}$$

For $t + \Delta t$, the impact responses such as plate and impactor displacements are determined by

$$\left\{\overline{F}\right\}^{t+\Delta t} = \{F\} + \{F_C^*\}^{t+\Delta t} + [M(\overline{\omega})](c_0\{\delta_P\}^t + c_1\{\dot{\delta}_P\}^t + c_2\{\ddot{\delta}_P\}^t) \tag{12}$$

$$\delta_I = \frac{\overline{F}_I^{t+\Delta t}}{m_I} \tag{13}$$

where,

$$\overline{F}_I^{t+\Delta t} = m_I(c_0\delta_I^t + c_1\dot{\delta}_I^t + c_2\ddot{\delta}_I^t) - F_C^{*t} \tag{14}$$

Plate and impactor's acceleration and velocity can be calculated as

$$F_{ix} = 0, \quad F_{iy} = 0, \quad F_{iz} = F_C^* \tag{15}$$

$$\{\ddot{\delta}_P\}^{t+\Delta t} = c_0(\{\delta_P\}^{t+\Delta t} - \{\delta_P\}^t) - c_1\{\dot{\delta}_P\}^t - c_2\{\ddot{\delta}_P\}^t \tag{16}$$

$$\ddot{\delta}_I^{t+\Delta t} = c_0(\delta_I^{t+\Delta t} - \delta_I^t) - c_1\dot{\delta}_I^t - c_2\ddot{\delta}_I^t \tag{17}$$

$$\{\dot{\delta}_P\}^{t+\Delta t} = \{\dot{\delta}_P\}^t + c_3\{\ddot{\delta}_P\}^t + c_4\{\ddot{\delta}_P\}^{t+\Delta t} \tag{18}$$

$$\dot{\delta}_I^{t+\Delta t} = \dot{\delta}_I^{t\Delta t} + c_3\ddot{\delta}_I^t + c_4\ddot{\delta}_I^{t+\Delta t} \tag{19}$$

At initial boundary condition

$$\{\delta_P\} = \{\dot{\delta}_P\} = \{\ddot{\delta}_P\} = 0 \tag{20}$$

$$\delta_I = \ddot{\delta}_I = 0 \text{ and } \dot{\delta}_I = V \tag{21}$$

The value of constants of integration can be obtained by

$$c_0 = \frac{1}{X\Delta t^2}, \quad c_1 = \frac{1}{X\Delta t}, \quad c_2 = \frac{1}{2X} - 1, \quad c_3 = (1-Y)\Delta t, \quad c_4 = Y\Delta t \tag{22}$$

where $X = 0.5$ and $Y = 0.25$. Here, we have considered an isoparametric quadratic element for the FE formulation having eight nodes (refer to Fig. 2). The expression for shape function $\hat{S}_i$ can be shown as

$$\hat{S}_i = \frac{(1+\vartheta\vartheta_i)(1+\varepsilon\varepsilon_i)(\varepsilon\varepsilon_i+\vartheta\vartheta_i-1)}{4} \quad (\text{for i} = 1, 2, 3, 4) \tag{23}$$

$$\hat{S}_i = \frac{(1-\vartheta^2)(1+\varepsilon\varepsilon_i)}{2} \quad (\text{for i} = 5, 7) \tag{24}$$

$$\hat{S}_i = \frac{(1-\varepsilon^2)(1+\vartheta\vartheta_i)}{2} \quad (\text{for i} = 6, 8) \tag{25}$$

where for element, the local natural coordinates are represented by ε and ϑ.

In the case of FGM, the material properties gradation takes place following different laws. Here, we have considered power law, which is given as [32]

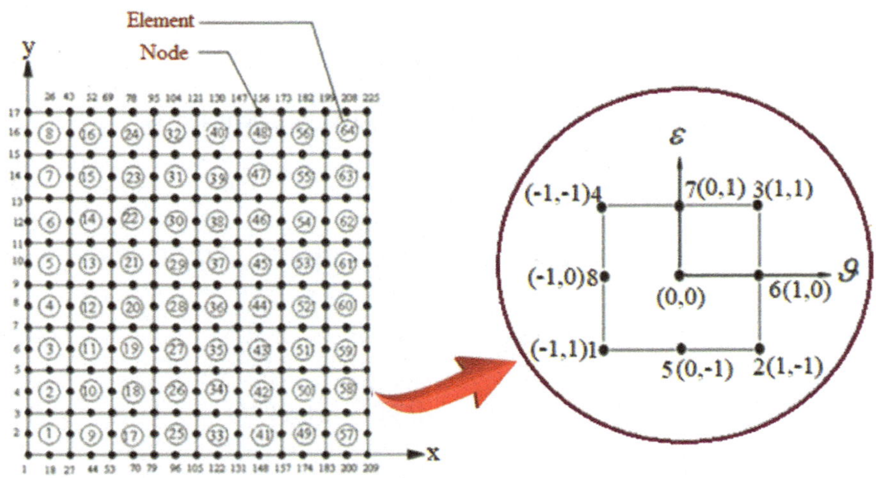

Fig. 2 Discretization of target plate into 8 × 8 mesh size

$$E(\hat{\omega}) = E_{Al} + (E_{Zr} - E_{Al})\left[\frac{z}{t} + \frac{1}{2}\right]^p \tag{26}$$

$$\mu(\hat{\omega}) = \mu_{Al} + (\mu_{Zr} - \mu_{Al})\left[\frac{z}{t} + \frac{1}{2}\right]^p \tag{27}$$

$$\rho(\hat{\omega}) = \rho_{Al} + (\rho_{Zr} - \rho_{Al})\left[\frac{z}{t} + \frac{1}{2}\right]^p \tag{28}$$

where E_{Al}, E_{Zr}, μ_{Al}, μ_{Zr}, and ρ_{Al}, ρ_{Zr} are the elastic modulus, Poisson's ratio, and mass density of aluminium and zirconia, respectively. 'p' shows the power-law index and 't' is the plate thickness, while $(\hat{\omega})$ represents the degree of stochasticity.

3 Radial Basis Function (RBF)

RBF model predicts the system responses for the structural analysis by considering the combination of all stochastic input parameters. An RBF network used radial function as activation function [33]. Hardy [34] first applied the multi-quadric (MQ) radial function to interpolate scatter data for geographical analysis where more than thirty functions are interpolated. Later on, Kansa [35] employed Hardy's MQ radial function for the partial differential solution of hydrodynamic problems. In the RBF, function responses are monotonically incremented or decremented from the central point with the radial distance [36]. The input pattern is given to the input layer, which connects the network to the environment. The input pattern is the m-dimensional vector, which the RBF function will classify.

RBF is the meshless computational approach to solve the problem expressed in the irregular domain. It is extensively applied in computer graphics, surface reconstruction, medical applications, fuzzy systems, pattern recognition, geosciences, and neural network because of its unique benefits such as superior convergence, meshless, and simplicity [37]. RBF is employed for large scatter data with m-dimensional where $m > 2$ which is based on the distance [38]. The distinctive feature of this model is that it does not require mesh generation; only geometrical properties are required to determine the distance between two pairs of points. It is easy to calculate the distance in both the low and high dimensional problems, so the RBF is much easier than the finite element analysis. The RBF model establishes the relationship between vectors of 'm' real-valued input features (variables) $y = (y_1, y_2, y_3, \ldots, y_m)$ and single value output variable 'z'. The surrogate model's objective is established using the finite number 'm' of input–output data sets (training observation/data cases) $\left(y_{(i)}, z_{(i)}\right)$, $i = 1, 2, 3, \ldots, m$ to predict the output for unknown input parameters. This data set $\left(y_{(i)}, z_{(i)}\right) \in D$ influences the hypothesis $f(x)$, which depends on the center point distance and $f(\|y - y_j, \gamma\|)$ the shape parameters. The RBFcentered at y_j are shown as [39]

$$f_j(y) = f(\|y - y_j\|) \in D^d, \quad \text{j} = 1, 2, 3, \ldots \text{d} \tag{29}$$

wherein domain *D*, the Euclidean distance is the $f(\|y - y_j\|) = \text{r}$ between pair of points. In standard form, RBF is given by Biazar and Hosami [40]

$$\begin{aligned} f_j(y) &\equiv e^{-\gamma^2 r^2} \quad \text{(For Gaussian spline)} \\ f_j(y) &\equiv \left(r^2 + \gamma^2\right)^{-\frac{1}{2}} \quad \text{(For inverse multi-quadric)} \\ f_j(y) &\equiv \sqrt{r^2 + \gamma^2} \quad \text{(For multi-quadric)} \\ f_j(y) &\equiv r^2 \log r \quad \text{(For thin plate)} \end{aligned} \tag{30}$$

The shape parameter $\gamma^2 = 1$ is assumed for the present study. It influences the precision of results, but it should be noted that true values in the sampling point are equal to the approximate function's functional values. After obtaining the RBF model outcomes, it is compared with the results of MCS for verification. The entire process followed for the present study is illustrated in Fig. 3.

4 RBF-Based Probabilistic Analysis of FGM Plates

In this chapter, the probabilistic input parameters taken are material properties, power-law exponent, temperature, and oblique impact angle. The effects of other input parameters are acknowledged as follows:

(a) Degree of stochasticity in material properties:

$$y_1\{E(\hat{\omega}), G(\hat{\omega}), \mu(\hat{\omega}), \rho(\hat{\omega})\} = \varphi[\{E(z)\}(\hat{\omega}), \{G(z)\}(\hat{\omega}), \{\rho(z)\}(\hat{\omega}), \{\mu(z)\}(\hat{\omega})]$$

(b) Power-law exponent:

$$y_2(p, \hat{\omega}) = \varphi[p, \{E(z)\}(\hat{\omega}), \{G(z)\}(\hat{\omega}), \{\rho(z)\}(\hat{\omega}), \{\mu(z)\}(\hat{\omega})]$$

(c) Temperature:

$$y_3(T, \hat{\omega}) = \varphi[T, \{E(z)\}(\hat{\omega}), \{G(z)\}(\hat{\omega}), \{\rho(z)\}(\hat{\omega}), \{\mu(z)\}(\hat{\omega})]$$

(d) Impact angle:

$$y_4(\theta, \hat{\omega}) = \varphi[\theta, \{E(z)\}(\hat{\omega}), \{G(z)\}(\hat{\omega}), \{\rho(z)\}(\hat{\omega}), \{\mu(z)\}(\hat{\omega})]$$

(e) Initial velocity of impactor:

$$y_5(V, \hat{\omega}) = \varphi[V, \{E(z)\}(\hat{\omega}), \{G(z)\}(\hat{\omega}), \{\rho(z)\}(\hat{\omega}), \{\mu(z)\}(\hat{\omega})]$$

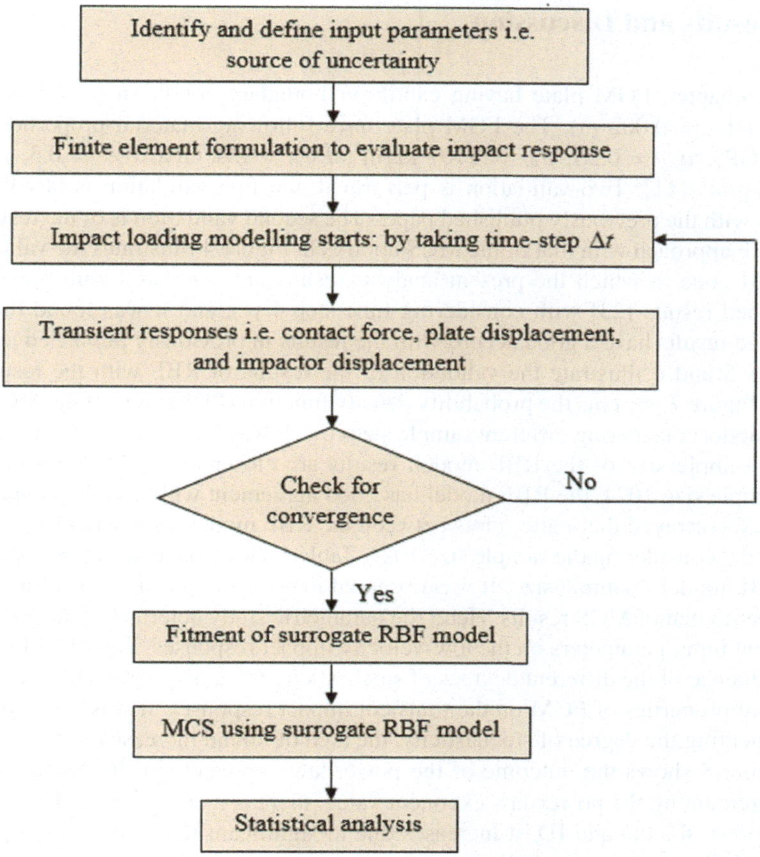

Fig. 3 Flowchart describing the present study's complete methodology

where φ represents the MCS operator for both combined and individual variation of input variables, while $\hat{\omega}$ is the degree of stochasticity. For simulating material properties' uncertainties-based stochastic approach, random variation in material properties with the predefined bond is considered. The considered variation occurs randomly in the limit of $\pm C\%$ concerning the deterministic value, where C represents the degree of stochasticity. For obtaining accurate outcomes from the original MCS, approximately 10^4 numbers of simulations are required, which is enormous. We aim to make this process time-efficient without negotiating with the results' accuracy. For this purpose, the RBF model is employed, which needs a significantly fewer number of simulations.

5 Results and Discussion

In this chapter, FGM plate having cantilever boundary condition ($l = 1$ m, $b = 1$ m, and $t = 0.002$ m). The FGM plate have following material properties, $E_{Al} = 70$ GPa, $\upsilon_{Al} = 0.25$, $\rho_{Al} = 2707$ kg/m^3, $E_{Zr} = 151$ GPa, $\upsilon_{Zr} = 0.3$, $\rho_{Zr} = 3000$ kg/m^3 [41]. Two validation is performed, the first validation is of FE code results with the previously published paper. The second validation is of the results of the RBF approach with that of the MCS approach. Figure 4 illustrates the validation of FEM code in which the present analysis results are compared with previously published results [23] with considering time step 1 μs, and it was found that the obtained results have a good accord with the results in previously published papers. Figures 5 and 6 illustrate the validation of the results of RBF with the results of MCS. Figure 7 presents the probability density function (PDF) plots of the MCS and RBF model considering different sample sizes (n). It was found that with an increase in the sample size of the RBF model, results are closer to the MCS results. For the sample size 1024, the RBF model has good agreement with the MCS approach. Figure 6 portrayed the scatter plots between the RBF model results and the original FE model considering the sample size 1024. Table 1 shows the convergence study of the RBF model's sample size. It was observed that for sample size 1024, the result is closer to that of MCS results. Here, the parametric study determined the effects of different input parameters on the low-velocity impact responses. Figure 7 illustrates the influence of the different degrees of stochasticity (C ± 5%, 10%, 20%) in all the material properties of FGM on the transient impact responses. It was found that on incrementing the degree of stochasticity, the area of bound increases.

Figure 8 shows the outcome of the power-law exponent (p). It was found that on incrementing the power-law exponent value, there is a decrement in CF's value. In contrast, for PD and ID, it increases due to significant reductions in the plate's stiffness. Figure 9 illustrates the outcome of temperature (T). It was found that the value of CF decreases with an increase in the temperature from 300 to 900 K, while

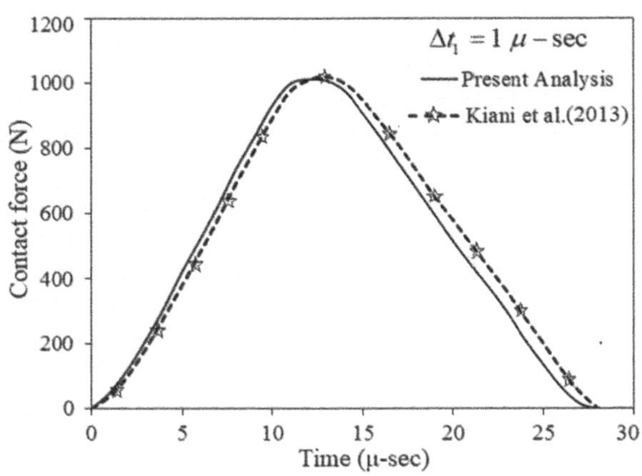

Fig. 4 Validation of present study by comparing time history of contact force of present study and previously published paper [11]

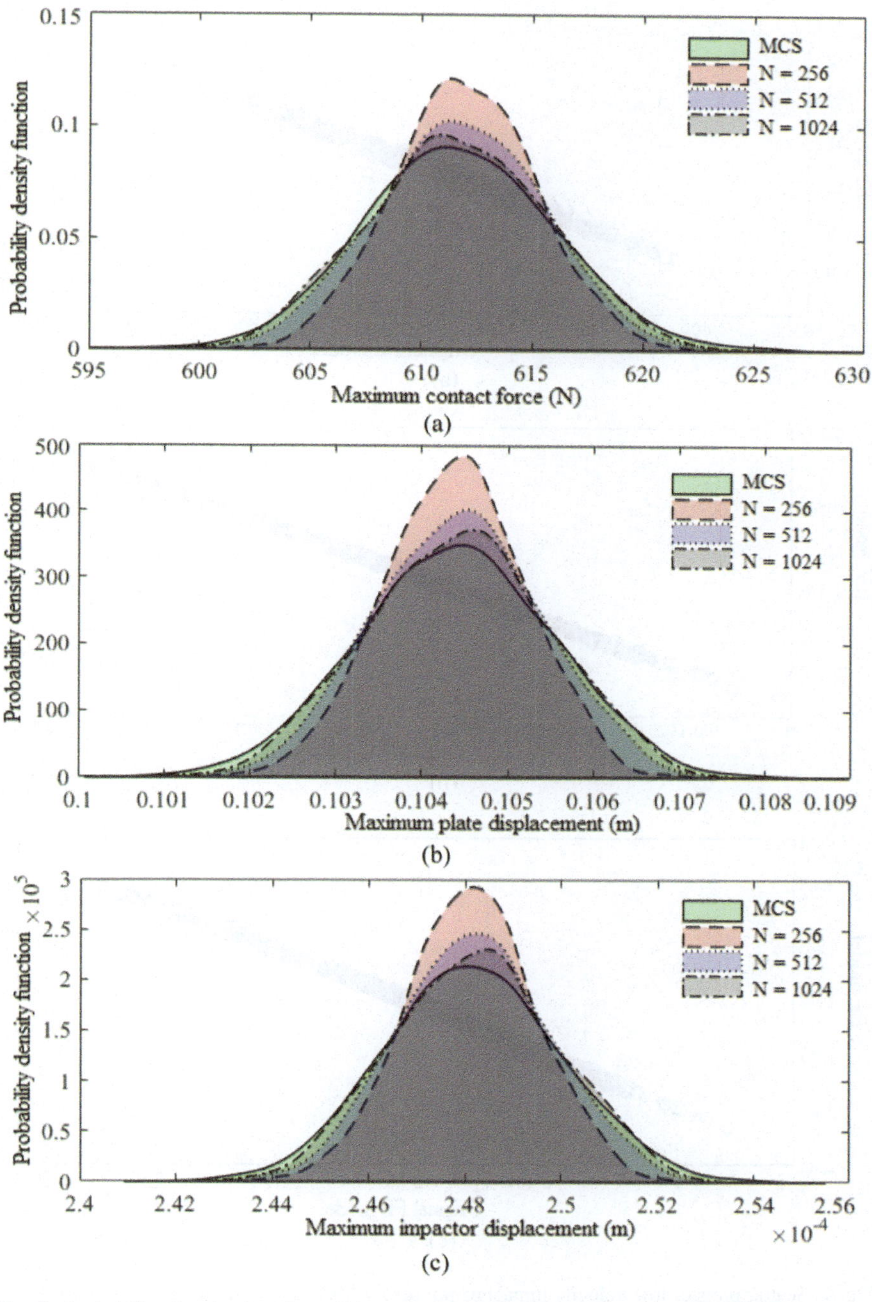

Fig. 5 Probability density function of low-velocity impact responses for MCS and RBF

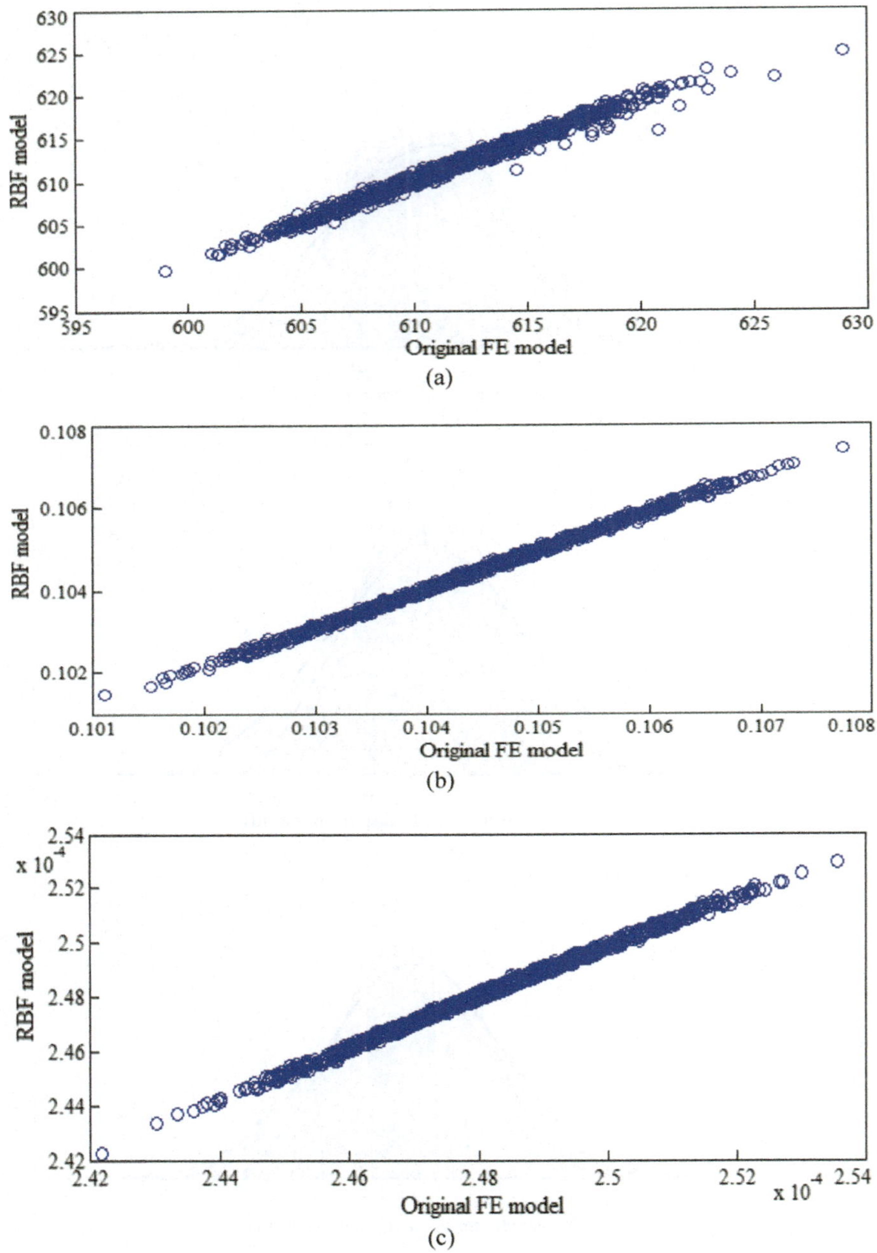

Fig. 6 Scatter plots of low-velocity impact responses for MCS and RBF model

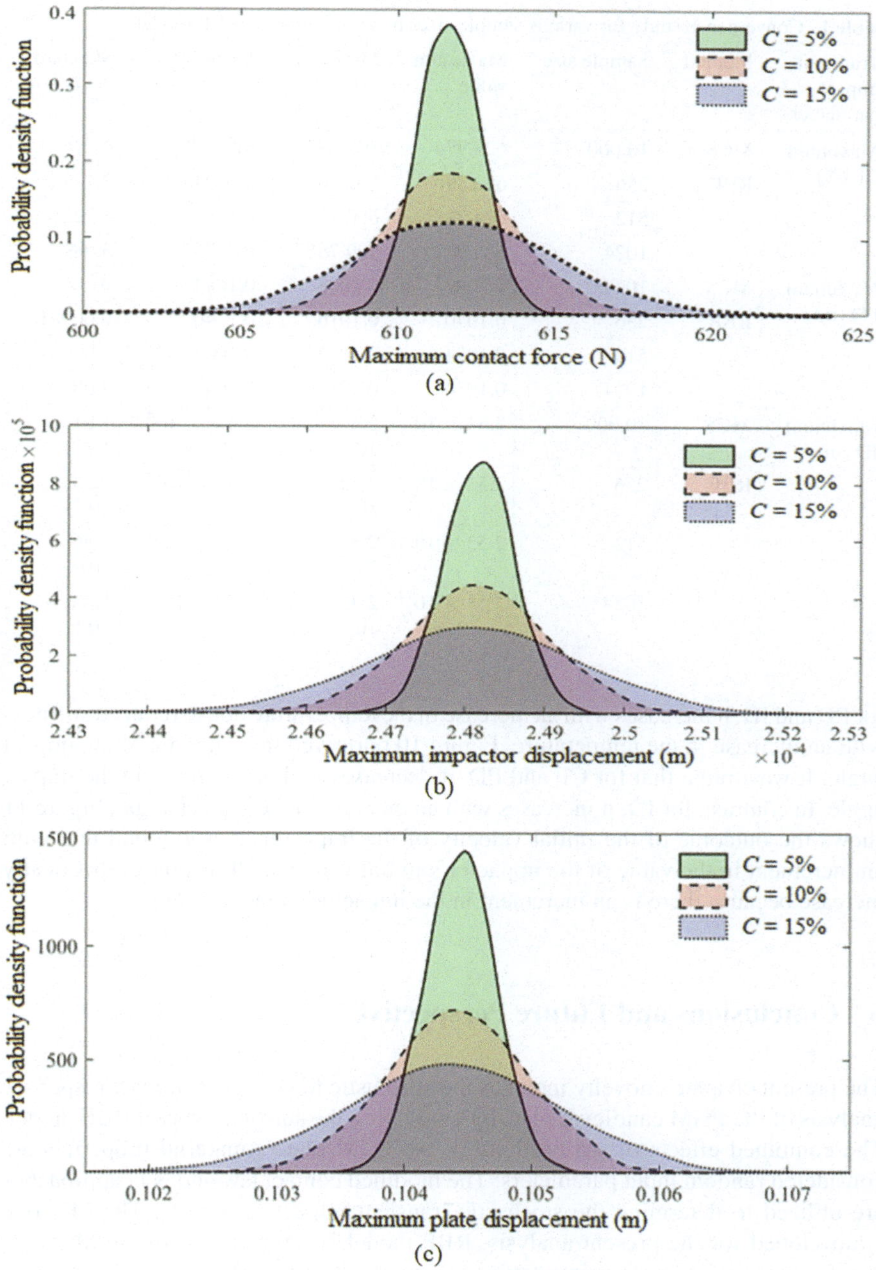

Fig. 7 Effect on stochastic transient impact responses due to change in the degree of stochasticity (C)

Table 1 Convergence study for various sample sizes by employing the RBF model

Transient impact parameter	Method	Sample size	Maximum value	Minimum value	Mean value	Standard deviation
Maximum CF (N)	MCS	10,000	628.974	597.697	611.771	4.276
	RBF	256	622.796	602.263	611.948	3.174
		512	624.229	600.813	611.852	3.722
		1024	625.032	599.785	611.757	3.989
Maximum PD (m)	MCS	10,000	0.1082	0.1008	0.1044	0.0011
	RBF	256	0.1070	0.1016	0.1040	0.00081
		512	0.1074	0.1012	0.1043	0.00096
		1024	0.1074	0.1010	0.1044	0.0010
Maximum ID (m)	MCS	10,000	2.54×10^{-4}	2.42×10^{-4}	2.48×10^{-4}	1.81×10^{-6}
	RBF	256	2.52×10^{-4}	2.43×10^{-4}	2.56×10^{-4}	1.33×10^{-6}
		512	2.53×10^{-4}	2.42×10^{-4}	2.52×10^{-4}	1.56×10^{-6}
		1024	2.54×10^{-4}	2.42×10^{-4}	2.48×10^{-4}	1.68×10^{-6}

for PD and ID, it increases with an increase in the temperature due to reduced stiffness with an increase in the temperature. Figure 10 portrayed the outcome of the impact angle. It was found that for CF and PD, it decreases with an increase in the impact angle. In contrast, for ID, it increases with an increase in the impact angle. Figure 11 shows the outcome of the initial velocity of the impactor. It was found that with an increment in the value of the impactor's initial velocity, all responses drastically increase because there is an increment in the impactor's kinetic energy.

6 Conclusions and Future Perspective

The present chapter's novelty includes the stochastic low-velocity impact responses analysis of the FGM cantilever plate by employing the surrogate-based RBF model. The combined effects of stochasticity in the FGM plate's material properties are considered random input parameters. The modified contact law and NTI approaches are utilized to determine the stochastic transient impact responses. The FE code is developed for the present analysis. RBF model is integrated with the MCS for constructing an efficient computational framework. The RBF model results with sample size 1024 have good agreement with MCS results in the convergence study. The parametric study finds the response of various input parameters on the stochastic transient low-velocity impact responses. The results illustrate that all the parameters

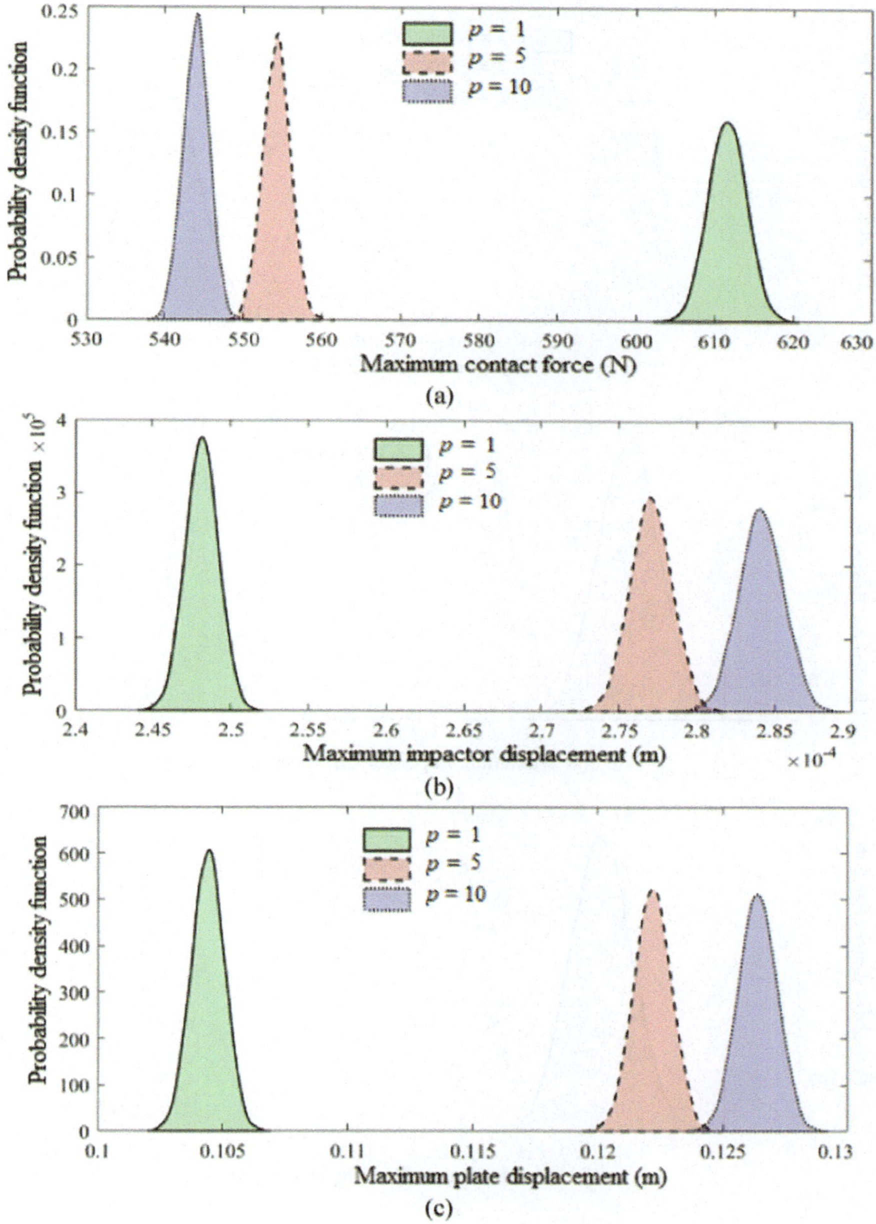

Fig. 8 Effect on stochastic transient impact responses due to change in power-law exponent (p)

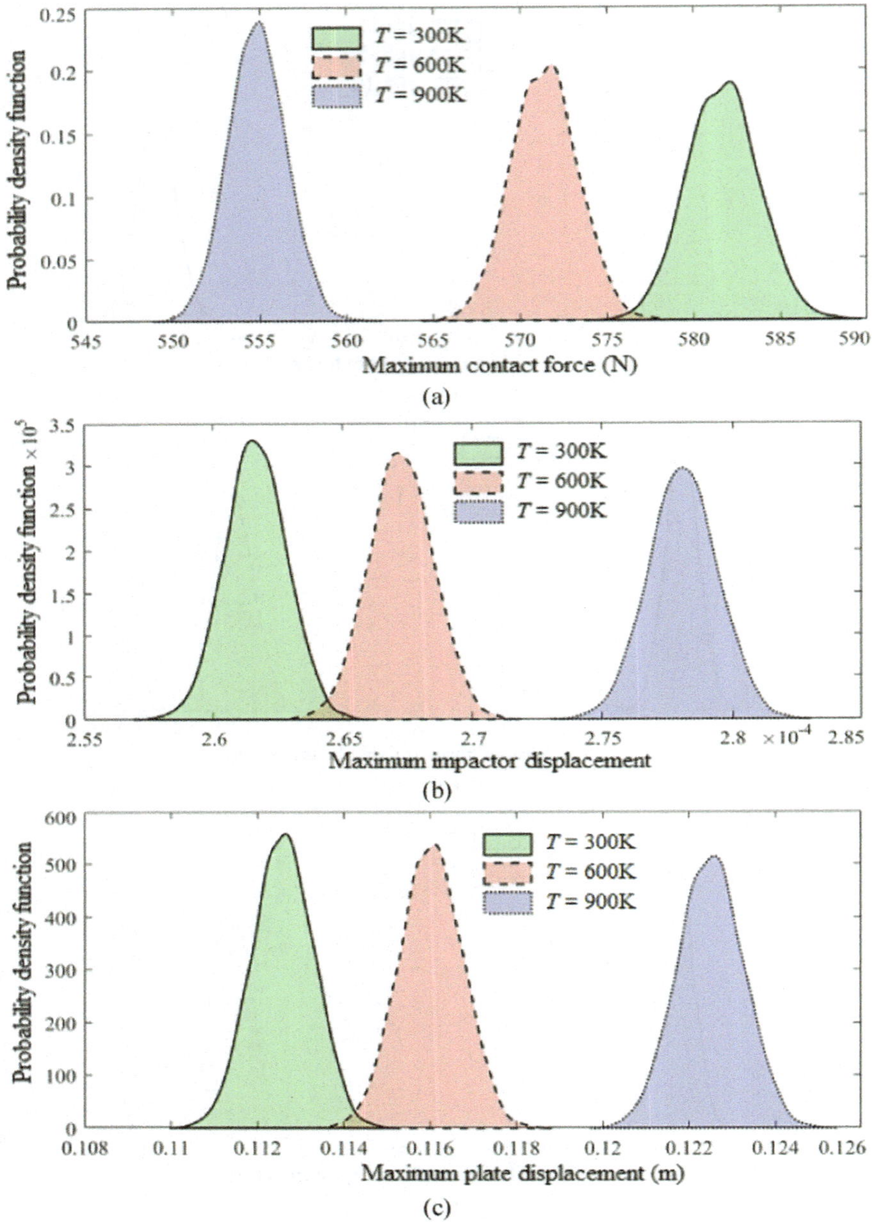

Fig. 9 Effect on stochastic transient impact responses due to change in temperature (T)

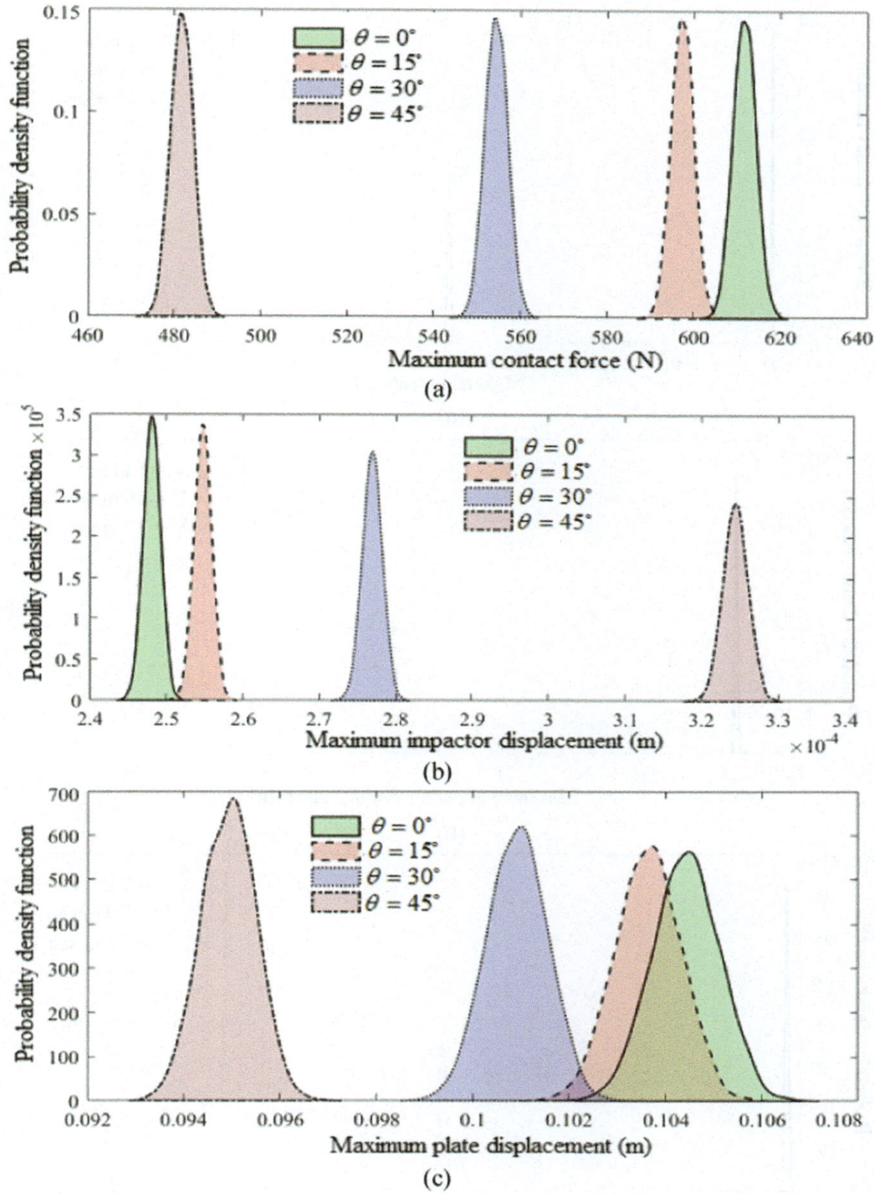

Fig. 10 Effect on stochastic transient impact responses due to change in impact angle (θ)

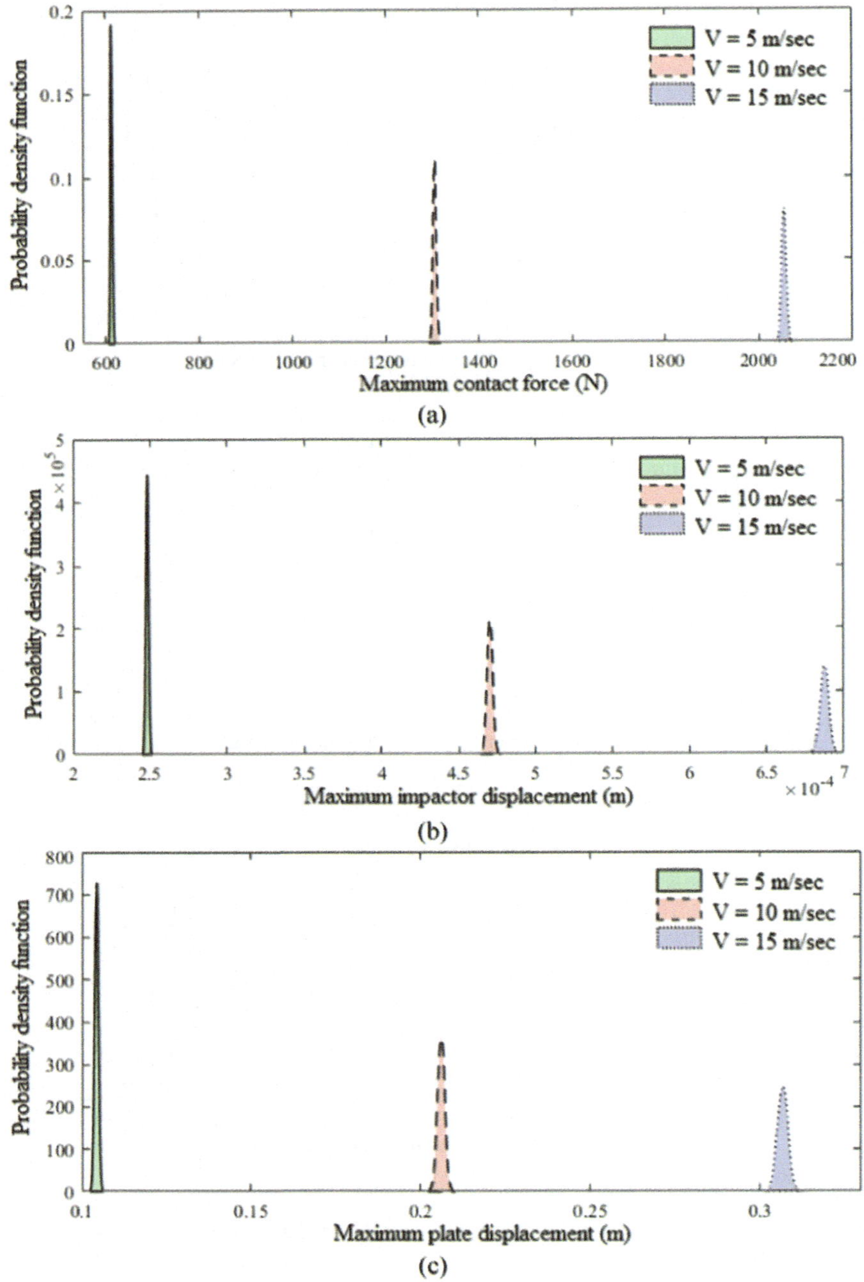

Fig. 11 Effect on stochastic transient impact responses due to change in initial velocity of the impactor (V)

such as stochasticity, power-law index, temperature, impact angle, and initial velocity of impactor significantly affect the impact responses. In the future, the employed framework can be utilized for the other dynamic analysis of layered structures.

Acknowledgements The third author received the final support from the MHRD, Govt. of India during the research period.

References

1. Zenkour AM (2005) A comprehensive analysis of functionally graded sandwich plates: part 2—buckling and free vibration. Int J Solids Struct 42(18–19):5243–5258
2. Reddy J (2000) Analysis of functionally graded plates. Int J Numer Meth Eng 47(1–3):663–684
3. Soto A, González EV, Maimí P, De La Escalera FM, De Aja JS, Alvarez E (2018) Low velocity impact and compression after impact simulation of thin ply laminates. Compos Part A: Appl Sci Manuf 109:413–427
4. Hsieh TH, Huang YS, Wang FX, Shen MY (2018) Impact and after-impact properties of nanocarbon aerogels reinforced epoxy/carbon fiber composite laminates. Compos Struct 206:828–838
5. Thorsson SI, Waas AM, Rassaian M (2018) Low-velocity impact predictions of composite laminates using a continuum shell based modeling approach part A: impact study. Int J Solids Struct 155:185–200
6. Schwab M, Todt M, Tauchner J, Schlie D, Pettermann HE (2018) Modeling, simulation, and experiments of high velocity impact on laminated composites. Compos Struct 205:42–48
7. Sun XC, Kawashita LF, Kaddour AS, Hiley MJ, Hallett SR (2018) Comparison of low velocity impact modelling techniques for thermoplastic and thermoset polymer composites. Compos Struct 203:659–671
8. Elamin M, Li B, Tan KT (2018) Impact damage of composite sandwich structures in arctic condition. Compos Struct 192:422–433
9. Karsh PK, Mukhopadhyay T, Dey S (2018) Stochastic investigation of natural frequency for functionally graded plates. In: IOP conference series: materials science and engineering, vol 326, no 1. IOP Publishing, UK, p 012003
10. Mao Y, Fu Y, Ai S, Fang D (2013) Interfacial damage analysis of shallow spherical shell with FGM coating under low velocity impact. Int J Mech Sci 71:30–40
11. Kiani Y, Sadighi M, Salami SJ, Eslami MR (2013) Low velocity impact response of thick FGM beams with general boundary conditions in thermal field. Compos Struct 104:293–303
12. Najafi F, Shojaeefard MH, Googarchin HS (2017) Nonlinear dynamic response of FGM beams with Winkler-Pasternak foundation subject to noncentral low velocity impact in thermal field. Compos Struct 167:132–143
13. Eghtesad A, Shafiei AR, Mahzoon M (2012) Study of dynamic behavior of ceramic–metal FGM under high velocity impact conditions using CSPM method. Appl Math Model 36(6):2724–2738
14. Bekhadda A, Bensaid I, Cheikh A, Kerboua B (2019) Static buckling and vibration analysis of continuously graded ceramic-metal beams using a refined higher order shear deformation theory. Multidiscipline Model Mater Struct 15(6):1152–1169
15. Kumar P, Srinivas J (2017) Free vibration, bending and buckling of a FG-CNT reinforced composite beam: comparative analysis with hybrid laminated composite beam. Multidiscipline Model Mater Struct 13(4):590–611
16. Parida S, Mohanty SC (2018) Vibration analysis of FG rotating plate using nonlinear-FEM. Multidiscip Model Mater Struct 15(1):26–49

17. Abed MS, Ahmed PS, Oleiwi JK, Fadhil BM (2020) Low velocity impact of Kevlar and ultra high molecular weight polyethylene (UHMWPE) reinforced epoxy composites. Multidiscipline Model Mater Struct 16(6):1617–1630
18. Qanber ASG, Alhusseini RSS, Al-Kasob BDH, Jasim MH, Ranjbar M (2020) Effect of the multiple projectile on the low-velocity impact response of CNTs reinforced beam. Multidiscipline Model Mater Struct 17(1):1–17
19. Wali M, Abdennadher M, Fakhfakh T, Haddar M (2011) Dynamic analysis of an elasto-plastic sandwich subjected to low velocity impact. Multidiscipline Model Mater Struct 7(2):184–206
20. Dey S, Adhikari S, Karmakar A (2015) Impact response of functionally graded conical shells. Lat Am J Solids Struct 12:133–152
21. Dey S, Karmakar A (2013) Dynamic analysis of delaminated composite conical shells under low velocity impact. J Reinf Plast Compos 32(6):380–392
22. Liu Y, Wang X, Wang L (2019) Interval uncertainty analysis for static response of structures using radial basis functions. Appl Math Model 69:425–440
23. Er PV, Tan KK (2018) Machine vibration analysis based on experimental modal analysis with radial basis functions. Measurement 128:45–54
24. Li X, Gong C, Gu L, Gao W, Jing Z, Su H (2018) A sequential surrogate method for reliability analysis based on radial basis function. Struct Saf 73:42–53
25. Wu Z, Wang W, Wang D, Zhao K, Zhang W (2019) Global sensitivity analysis using orthogonal augmented radial basis function. Reliab Eng Syst Saf 185:291–302
26. Karsh PK, Mukhopadhyay T, Dey S (2019) Stochastic low-velocity impact on functionally graded plates: probabilistic and non-probabilistic uncertainty quantification. Compos B Eng 159:461–480
27. Karsh PK, Kumar RR, Dey S (2019) Stochastic impact responses analysis of functionally graded plates. J Braz Soc Mech Sci Eng 41(11):1–13
28. Dey S, Karmakar A (2012) Free vibration analyses of multiple delaminated angle-ply composite conical shells—a finite element approach. Compos Struct 94(7):2188–2196
29. Sun CT, Chen JK (1985) On the impact of initially stressed composite laminates. J Compos Mater 19(6):490–504
30. Shariyat M, Nasab FF (2014) Low-velocity impact analysis of the hierarchical viscoelastic FGM plates, using an explicit shear-bending decomposition theory and the new DQ method. Compos Struct 113:63–73
31. Bathe KJ (2006) Finite element procedures. Prentice Hall Publication
32. Mukhopadhyay T, Naskar S, Chakraborty S, Karsh PK, Choudhury R, Dey S (2021) Stochastic oblique impact on composite laminates: a concise review and characterization of the essence of hybrid machine learning algorithms. Arch Comput Methods Eng 28(3):1731–1760
33. Sun G, Sun Y, Wang S (2015) Artificial neural network based inverse design: airfoils and wings. Aerosp Sci Technol 42:415–428
34. Hardy RL (1990) Theory and applications of the multiquadric-biharmonic method 20 years of discovery 1968–1988. Comput Math Appl 19(8–9):163–208
35. Kansa EJ (1985) Application of Hardy's multiquadric interpolation to hydrodynamics (No. UCRL-93522; CONF-860119-2). Lawrence Livermore National Lab, CA
36. Queipo NV, Haftka RT, Shyy W, Goel T, Vaidyanathan R, Kevin-Tucker P (2005) Surrogate-based analysis and optimization. Prog Aerosp Sci 41(1)
37. Li Y, Hon YC (2017) Finite integration method with radial basis function for solving stiff problems. Eng Anal Boundary Elem 82:32–42
38. Majdisova Z, Skala V (2017) Radial basis function approximations: comparison and applications. Appl Math Model 51:728–743

39. Naskar S, Mukhopadhyay T, Sriramula S, Adhikari S (2017) Stochastic natural frequency analysis of damaged thin-walled laminated composite beams with uncertainty in micromechanical properties. Compos Struct 160:312–334
40. Biazar J, Hosami M (2017) An interval for the shape parameter in radial basis function approximation. Appl Math Comput 315:131–149
41. Singh H, Hazarika BC, Dey S (2017) Low velocity impact responses of functionally graded plates. Procedia Eng 173:264–270

Application of Machine Learning in Determining the Mechanical Properties of Materials

Naman Jain, Akarsh Verma, Shigenobu Ogata, M. R. Sanjay, and Suchart Siengchin

1 Introduction

As per the national research councils of India material characterization should describe feature such as structure of material, composition, defects, etc. with application. Material characterization involves chemical, physical and mechanical properties which describe the material in best manner. To understand the physics of engineering material under the action of external forces (which may be of contact and non-contact type) is known as mechanical characterization of material. These mechanical properties play an important role in characterization of material which helps the research to find its correct application. Evaluation of mechanical properties relies on experimental data [1] obtained from distractive and non-distractive testing. And mathematical model (based on governing law) which interpretiate the mechanical properties from data obtained from experiment. Increasing demand of the society result in discovering of new material having superior properties with vast range of application are been required. Evaluation in material science field can be best described into four phases or paradigms [2]. First phase start from stone-age past back to thousands of years where by only metallurgical observation help in predicting

N. Jain
Department of Mechanical Engineering, ABES Engineering College, Ghaziabad, India

A. Verma (✉) · S. Ogata
Department of Mechanical Science and Bioengineering, Graduate School of Engineering Science, Osaka University, Osaka 560-8531, Japan
e-mail: akarshverma007@gmail.com

A. Verma
Department of Mechanical Engineering, University of Petroleum and Energy Studies, Dehradun, India

M. R. Sanjay · S. Siengchin
The Sirindhorn International Thai-German Graduate School of Engineering, King Mongkut's University of Technology North Bangkok, Bangkok, Thailand

V. Kushvaha et al. (eds.), *Machine Learning Applied to Composite Materials*, Composites Science and Technology, https://doi.org/10.1007/978-981-19-6278-3_5

the characteristic of material, this is purely empirical. Second phase start near about sixteenth century where scientists have developed mathematical models to correlate the physical relations. In these paradigms conventional laws have been formulated in the form of mathematical equation. But with time the complexity of developed mathematical model become too complex, result in raising question on the solution obtain from the above analytical methods. In the mid of nineteenth century, advancements in computational science increased that gave rise to the third phase. Any mechanical properties of the material cannot be best described by only one mathematical model or conventional law; there may be several concept which may describe the single property in best way. Therefore experimental data is been required which is been interpreted through mathematical models with the help of computational science. Molecular dynamic and density function theory are been some of the examples in material science in this paradigm. Data generated from experiment and simulation from computer give rise to fourth phase where calculation of mechanical properties is based on large set of data also known as data driven paradigm [3]. Large data set obtained from different experiments including structure patterns help in determining the hidden correlation which were not visible in small data sets result in horizon of twenty-first century research. Figure 1 represents the four paradigms of material science starting from stone-age to twenty-first century.

2 Big-Data

To meet the specific requirement of industries or scientist discovery of novel material or unknown properties of known materials is been evaluated on the basic of data obtained from different source which may or may not relate the property in theoretical concepts. But contain some hidden correlation that may describe the behavior of the material and the huge amount of data obtained from different source comes under big-data. Big-data is been characterized by five-V models i.e. *volume*: concern about the size of the data, *velocity*: data generation speed, *variety*: diversity of data generated, *veracity*: quality and accuracy of data generated and *value*: value provided the data generated [4, 5]. Development in sensors helps in collection of data, but bigness of data have their own challenges such as storage, visualization, analysis, and retrieval. Suppose data obtained from heterogeneous sources (from different experiments) have the problem of correlation in between them, inconsistencies within it, some missing value, reliability issue, etc. These are the major challenges comes in the path of scientist while working in the field of material science. In the field of computational material science, European center of excellence, Novel Material discovery (NOMAD) having concept "FAIR" of big-data i.e. **F**indable for anyone interest, easily **A**ccessible, standard representation which make it **I**nteroperable and **R**eusable [6]. Different stages involve in mechanical characterization of material are and also shown in Fig. 2, data obtained from the heterogeneous source may first pre-process to identify the correlation of in between the data; after that identification of key feature/factor which may help in data reduction/selection; after that mathematical

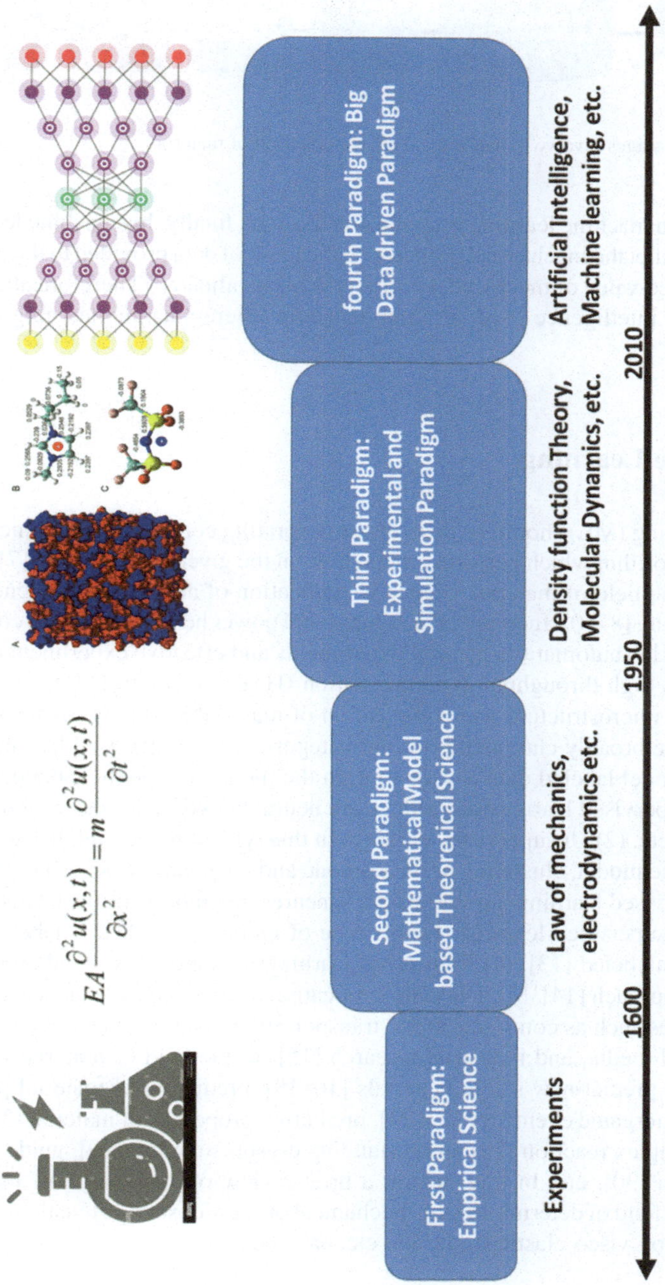

Fig. 1 Four paradigms of material science starting from stone-age to twenty-first century

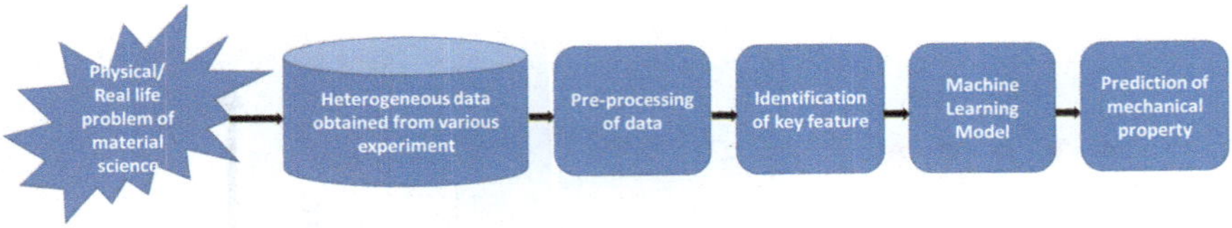

Fig. 2 Different stages involve in mechanical characterization of material

model based on machine learning is applied on the data; finally this machine learning model will predict the mechanical property on the basic of data provided to the model. At last predicted value of mechanical property is been validated. Therefore after big-data computer intelligence plays an important role where machine learning comes into picture.

3 Machine Learning

Machine learning (ML) should be describe as generating computer intelligence with the help of algorithm which perform certain task on the given relevant data [7]. Next paradigm in the field of material science is application of artificial intelligence and machine learning [8–10]. Increase in computational power help the scientific community to involve lab automation, parallel experiments and effective experiment design also known as high throughput experimentation (HTE). Through HTE correlation in processing, microstructure and composition of material [8–11] is been possible. ML models are broadly classified into four categories: (1) Supervised Learning: In this type of model labeled data is used to train the model and predicts the outcome [12]. Some supervised learning algorithms are neural networking, linear regression, Naive Bayes, etc. (2) Unsupervised learning: In this type of model unlabeled data is used to train the model, clustering of data is done and the outcome is been predicted. Some unsupervised learning algorithms are k-nearest neighbors, k-mean clustering, etc. (3) Semi-supervised learning: In this type of model some data is labeled data and some is unlabeled [13]. (4) Reinforced learning: this model is based on reward and penalty approach [14]. ML has wide application in every area of science and day to day activities such as communication, transportation, medical, business, material science, social media, and industrial research [15]. In the field of material science ML is used to predict new stable materials [16–18], predicting of material properties [19–21], inorganic chemistry [22–24], predicting properties of material [25–27], analyzing complex reaction [28], understanding crystal structure [29], guide experimental design [30], etc. In this chapter a brief review of application of machine learning in the field of determining the mechanical properties such as tensile strength, fatigue behavior, visco-elastic study, and etc. have be done.

4 Tensile Strength

Characterization of material is done through determining the mechanical properties such as tensile strength, yield strength, % elongation, etc. It is the behavior of material under the action of external forces. Mechanical properties of material depend upon many factors such as microstructure, alloying composition, fabrication processing parameters, and other external factors. As per present scenario most of the researcher employ the universal testing machine to determine the mechanical properties of material. But there is no mathematical model where all the above governing factors will be include, therefore the role of machine learning comes into picture. Wang et al. [31] applied the machine learning algorithm to determine the elongation and yield strength of reduce activation ferritic/martensitic steels (RAFM). To perform the ML algorithm database is been developed through experiments. And to maintain dimensional equality normalization was performed as preprocessing method for feature engineering. About 80% database is been used to train the ML model and 20% to test. Initially there are about 19 feature, after artificial selection 8 feature (due to narrow data range following were content Ni, Nb, V, Mn, Mo, neglected; due to testing accuracy following were neglected content of P, S; and due to small dataset of time, temperature of austenitizing is been neglected) were ignored. Random forests model was employed as ML algorithm to find out the impact of remaining 11 features (content of C, Cr, W, Si, V, Ta, Ti, N, B, Temp., and Time) on yield strength and elongation of RAFM. Moreover, last-place elimination rule was also been employed to eliminate the feature which less correlate. Results show that tempering temperature, content of C and tempering time show the highest correlation in predicting the yield strength of RAFM followed by content of Cr, Ti, W and B. Outcomes of ML model shows similarity with RAFM steels principle of physical metallurgy i.e. formation of MX carbonitride during tempering process of RAFM steel play an important role in strengthening mechanism [32].

Sasikumar et al. [33] predict the ultimate tensile strength of epoxy/unidirectional carbon composites using back propagation algorithm based on Artificial Neural Network (ANN). Acoustic Emission (AE) response of epoxy/unidirectional carbon fiber is been recorded, AE sensors were placed on both side of the specimen. Total of about 18 dataset were generated and three mode of failures were studied i.e. (1) fiber breakage: since most of the load is taken up by the fiber, therefore highest energy and amplitude is been generated, (2) delamination (in case of unidirectional fiber splitting of fiber): duration of this mode is short and amplitude and energy is also lesser than fiber breakage failure and (3) matrix cracking: this failure occurs throughout the test. AE characteristic produce were recorded for each failure mode i.e. AE hits and ultimate tensile strength. Three layers ANN model were used to fit the data obtained with 45 neurons at middle layer, Levenberg-Marguart algorithm employed with linear transfer function. Results show about 9.5% error which makes ANN more provident method to predict the tensile strength using AE emission.

Santos et al. [34] apply machine learning approach to predict the mechanical properties with foundry process parameter. 25 variable parameters were taken into account for 889 casting datasets. About 10 different ML classifiers such as Bayes TAN, Bayes Hill Climber, Bayes K2, Naive Bayes, KNN algorithms were used. Results show that Hill Climbing classifier give best result for lower training datasets whereas in case other classifier their accuracy increases with increase in training dataset.

Sterjovski et al. [35] apply the artificial neural networks to predict the mechanical properties of steels. Focus of the work is on to predict the impact strength of quenched and tempered pressure vessel steel exposed to multiple postweld heat treatment cycles on the basic of alloy content and heat treatment processing parameter (Model 1); hardness of the simulated heat affected zone of weldment in pipeline and tap fitting steels after in service welding on the basic of alloy contents and cooling time (Model 2); and the hot ductility and hot strength of various micro alloyed steels over the temperature range for strand or slab straightening in the continuous casting process on the basic of treatment conditions and chemical composition (Model 3). ANNs model variables is classified into material characteristics (composition of alloying element, such as %C, %Mn, %Cr, etc., thickness and hardness) and processing parameters (temperature, cooling rate, time, etc.). For Model 1 10 input variable i.e. cooling type, time, thickness, orientation, temperature, content of C, Cr, Mn, and B were studied. In Model 2 16 parameters were taken and 9 were taken for Mode 3. For all the models content of carbon play a significant impact. For Model 1 holding time and temperature play important role, impact strength decreases with PWHT time. In Model 2 with decrease in cooling rate from 150 to 30 °C/s result in drop in hardness. For Mode 3 with increase in test temperature hot tensile strength decreases.

Datta et al. [36] apply the concept of neural networks and genetic algorithms in designing multiphase steel for balanced strength and ductility. Major challenges in multiphase steel to use the optimum amount of alloying elements such that it has maximum strength without compromising the ductility. Moreover, thermo-mechanical controlled processing parameters such as cooling rate, rolling temperature, reheating temperature and deformation in rolling stages also play an important role. For modeling of neural networking thermo-mechanical controlled processing parameters and alloy composition were taken as input variable and yield strength, elongation and tensile strength as output parameter. Further to reduce the neural network connectivity two methods were employed. First is intuitive pruning algorithm [37] which is used to remove the lesser significant connections. In second approach genetic algorithm [38, 39] (predator prey algorithm and pareto front is employed to minimize the training error). Results show that rolling temperature does not play any significant role in predicting the yield strength and ultimate tensile strength of multiphase steel. Reason regarding that conclusion is that rolling temperature is maintain higher that the recrystallization temperature therefore amount of dislocation during rolling is lesser due to which alteration in yield strength and ultimate tensile strength is insignificant. Major factor which influence the multiphase steel are precipitation hardening and post rolling cooling rate. Results show that carbide forming elements have less effect as compare to Cu (strain hardening) in case ultimate tensile strength. Whereas cooling rate after rolling play important role in microstructure construction and found to most significant input variable even for percentage elongation also.

Zhang et al. [40] applied the deep learning concept for elasticity imaging of non-homogeneous material through Physics-Informed Neural Networks (PINNs) on the basic of which diseased and normal tissue cab be identified. Accomplishment of objective was done using two neural network one for solution approximation and other for unknown material estimation. The objective is to determine the material properties by observing displacement data for plain strain problems for incompressible Neo-Hookean hyper-elastic materials. Formulation loss function involves displacement data, physics information of partial differential equation, boundary condition and incompressibility constraint.

Matzbower et al. [41] applied the concept of neural networking to approximate the ductility and strength of welding shipbuilding steel alloys. Database of 189 weld steel alloys samples were been generated which include chemical composition (such as content of C, Mn, Si, Cr, Ni, Mo, Cu, S, P etc.), elongation and cross-section area reduction and weld cooling rate as an input variable and ultimate tensile strength and yield strength as output variable. All input variable were been normalized within the range of $\pm$ 0.5. About 80 neural models were created with varied number of hidden units and 5 random initial seed set. The output y is the linear function of bias and hyperbolic tangent transfer function. Results show that about ¼ dataset doesn't satisfied the model in case yield strength. Noteworthy input variable which had significant impact on yield strength are content of Mn, C, Ni, Ti, Mo and cooling rate.

Shigemori et al. [42] applied the concept of Just-In-Time (JIT) methodology to developed quality design system for steel. Focus of the work is to decide the best manufacturing procedure as per the desire quality target for steel bar. Initially the production of steel slabs is done by altering the chemical composition. To obtained desire shape and size cold working (rolling process) of the steel slab heated to specific temperature is done. Desirable mechanical properties and microstructure is been obtained through cooling process which may further be temper rolled and heat treated to again modified the properties. In this work quality index of the steel or output parameter taken were tensile strength, yield strength, toughness and elongation. And on the other hand alloying element content, cooling process, rolling process and extraction temperature were taken as input parameter for mathematical model. Mathematical model is based on locally weighted regression approach type of JIT modeling which can be employed for nonlinear processes. Quality designing system consists of three models first one to select the initial input values so that value of initial objective function (it is deviation of product quality from target value) is minimum; second one for locally weighted regression models which generate nonlinear relation between quality and manufacturing condition, and third module is for deriving the optimal manufacturing conditions by employing quadratic programing methods. Results show that accuracy of locally weighted regression is higher as compared linear regression.

Swaddiwudhipong et al. [43] employed the least square support vector machines to characterize the aluminum alloys (Al6061 and Al7075). Previously many researcher had used reverse analysis through iterative process but due to high nonlinear nature make it complex. Other approach is to use analytical model developed between load-indentation parameters and material properties. Latter on concept of artificial neural networking is also been adopted by the researchers [44, 45] for material characterization when single indenter is been used to extract the data for thin films, which further extended to elasto-plastic materials [46]. In present work validated results obtained from FEA (using ABAQUS software) of indentation tests using conical indentor based on large strain and deformation model were taken as input in least square support vector machines (LS-SVM). The four LS-SVM models were developed using MATLAB 6.5 based on structural risk minimization principle. Problems like over fitting and large deviation are overcome by regularization and final unction is obtained with incorporation of regularization and loss function. Outcome shows that LS-SVM model gives more accurate result of Young's modulus and yield strength rather than position's ration.

5 Fatigue and Creep

Behavior of material under the action of cyclic loading is known as fatigue. Fatigue behavior/characteristic of material is determined by calculating the no. of cycles before failure also known as fatigue life. Abdalla and Hawileh [47] employed radial basis function (RBF) ANN for modelling and simulation of fatigue life (low-cycle fatigue life) of steel bar. For training the ANN model input parameters were maximum tensile strain (sinusoidal axial strain with 0.05 Hz frequency ranging from 3 to 10% beyond yield point) and strain ratio (−1, −0.75, −0.5, −0.25 and 0). And no. of cycles to fail the material is been evaluated through fatigue test taken as output parameter which is been evaluated by Coffin-Manson equation [48–50]. ANN model is based on back-propagation feed forward with random selection of initial weight values. ANN model consist of two additional hidden layers with four neurons in each layer. First hidden layer consist of Gaussian activation function while one output layer and two additional hidden layers consist of tangent hyperbolic activation function. Accuracy of ANN model is high with the normalized square error 0.0428 and R value 0.9869.

Application ANN also been used predict the failure life of composites material now a days. Lee et al. [51] predict the fatigue life of the carbon and glass fiber reinforced epoxy composites using ANN method. Object of the work is to develop the model for predicting the fatigue life for newly developed/existing material having lesser fatigue testing results. Only approach available is developed mathematical model using physical laws by evaluating material properties and damage at microstructure which is time consuming and yet satisfactory result is also not obtained. Dataset is been obtained by fatigue testing of material with R rations 0.1–0.5. Three input parameters were taken i.e. are stress ratio, stress range and mean stress and on the other hand fatigue life was taken as output parameters. ANN model prediction give good result for median life as compared to minimum failure life.

Agarwal et al. [52] determine the fatigue strength of steel based on composition and processing parameter by employing different data science techniques such as ANN, decision trees and regression. Database is been obtained from MatNavi (National Institute of Material Science) which is world largest database developed at Japan. Database consists of low alloy steels, spring steels and carburizing steels having 25 input variable parameters having following classifications: (1) Chemical content such as weight percentage of different alloying element C, Si, Mn, S, Ni, Cr, Mo, Cu, etc.; (2) Upstream processing parameters i.e. reduction ratio, ingot size, non-metallic inclusions, etc.; (3) Heat treatment parameters such as time, temperature, and processing conditions for different process, and; (4) Mechanical properties such as yield strength, ultimate tensile strength, %elongation, hardness, fatigue strength, impact value, etc. Preprocessing of database is done for feature selection and R, WEKA and MATLAB were used as data processing software. Result shows that tempering time, Carburization temperature and Diffusion temperature have the highest correlation with fatigue strength. About 12 different ML models were used out of which MPR and M5 model tress show the best result with R^2 value of 0.98.

Zhang et al. [53] applied the concepts of genetic algorithm and support vector machine (SVM) to determine the fatigue life of material. SVM machine learning technique is applied when the database contain smaller sample with good accuracy result. In this technique nonlinear mapping of input variable into infinite-dimension space by creating hyper plane which classify the database into classes. Kernel function (Gauss function) was also been incorporated to solve nonlinear problem through SVM. Purpose for introducing the kernel function is to overcome the dimensionality problem, help in nonlinear transformation expressions and provide different mapping space having different properties. Accuracy of SVM depends upon the three factors: Penalty factor, higher is the value higher is the fitting but time consuming also even too high value result in over-fitting; Lower fitting accuracy, higher is value lower the no. of support vectors result is reduction in the complexity of the accuracy of model; γ, smaller is the value, poor performance of radial based function and vice versa but too high value may cause poor generalization ability. Genetic algorithm approach was employed to optimize the parameters as parameter selection is one of the major problem where theoretical guidance cannot help. SVM model was used to predict the P91 steel base metal and welding consumable taken from literature of Ji [54]. Holding time is taken as input parameter whereas fatigue life as output parameter. Genetic algorithm-SVR shows the highest accuracy with R2 of 0.99 on training sample. The authors have a vast experience in the field of molecular dynamics and experimental characterization of composites materials [55–106].

6 Conclusion

This chapter shows the various machine learning techniques applied by different researchers towards characterization of mechanical properties of material. These articles review literature on the application of machine learning in the field of fatigue failure, creep-recovery behavior of material, tensile strength, etc. Utilization of machine learning techniques in the field of material characterization is growing day by day. ML has a vast future in the field of material characterization which is not only limited to above characterization techniques.

Acknowledgements The authors are grateful to the monetary support provided by the University of Petroleum and Energy Studies (UPES)-SEED Grant program.

Conflicts of Interest "There are no conflicts of interest to declare by the authors."

References

1. Agrawal, Choudhary A (2016) Perspective: materials informatics and big data: realization of the "fourth paradigm" of science in materials science. Appl Mater 4:053208
2. Stoll A, Benner P (2021) Machine learning for material characterization with an application for predicting mechanical properties. GAMM-Mitteilungen 44:e202100003. https://doi.org/10.1002/gamm.202100003
3. Hey T, Tansley S, Tolle K (2009) The fourth paradigm: data-intensive scientific discovery. Microsoft Research
4. Chen CLP, Zhang C-Y (2014) Data-Intensive applications, challenges, techniques and technologies: a survey on big data. Inf Sci (Ny) 275:314–347. https://doi.org/10.1016/j.ins.2014.01.015
5. Lusher SJ, McGuire R, van Schaik RC, Nicholson CD, de Vlieg J (2014) Data-driven medicinal chemistry in the era of big data. Drug Discov Today 19(7):859–868. https://doi.org/10.1016/j.drudis.2013.12.004
6. Draxl C, Scheffler M (2018) NOMAD: the FAIR concept for big data-driven materials science. MRS Bull 43:676–682
7. Pilania G (2021) Machine learning in materials science: from explainable predictions to autonomous design. Comput Mater Sci 193:110360. https://doi.org/10.1016/j.commatsci.2021.110360
8. Long C, Hattrick-Simpers J, Murakami M, Srivastava R, Takeuchi I, Karen VL, Li X (2007) Rapid structural mapping of ternary metallic alloy systems using the combinatorial approach and cluster analysis. Rev Sci Instrum 78:072217
9. Long C, Bunker D, Li X, Karen V, Takeuchi I (2009) Rapid identification of structural phases in combinatorial thin-film libraries using x-ray diffraction and non-negative matrix factorization. Rev Sci Instrum 80:103902
10. Kusne AG et al (2014) On-the-fly machine-learning for high-throughput experiments: search for rare-earth-free permanent magnets. Sci Rep 4:6367
11. Suram SK et al (2016) Automated phase mapping with AgileFD and its application to light absorber discovery in the V–Mn–Nb oxide system. ACS Comb Sci 19:37–46
12. Alpaydin E (2014) Introduction to machine learning. The MIT Press, Cambridge, MA
13. Nguyen H, Maeda S, Oono K (2017) Semi-supervised learning of hierarchical representations of molecules using neural message passing. Preprint at arXiv:1711.10168

14. Sutton RS, Barto AG (2018) Reinforcement learning. The MIT Press, Cambridge, MA
15. Larranaga P, Atienza D, Diaz-Rozo J, Ogbechie A, Puerto-Santana CE, Bielza C (2018) Industrial applications of machine learning. CRC Press, Boca Raton
16. Lookman T, Eidenbenz S, Alexander F, Barnes C (eds) (2018) Materials discovery and design by means of data science and optimal learning. Springer International Publishing, Basel
17. Ryan K, Lengyel J, Shatruk M (2018) Crystal structure prediction via deep learning. J Am Chem Soc 140:10158–10168
18. Nouira A, Sokolovska N, Crivello J-C (2018) Crystalgan: learning to discover crystallographic structures with generative adversarial networks. Preprint at arXiv:1810.11203
19. Zheng X, Zheng P, Zhang R-Z (2018) Machine learning material properties from the periodic table using convolutional neural networks. Chem Sci 9:8426–8432
20. Carrete J, Li W, Mingo N, Wang S, Curtarolo S (2014) Finding unprecedentedly low-thermal-conductivity half-Heusler semiconductors via high-throughput materials modeling. Phys Rev X 4:011019
21. Kim C, Pilania G, Ramprasad R (2016) From organized high-throughput data to phenomenological theory using machine learning: the example of dielectric breakdown. Chem Mater 28:1304–1311
22. Isayev O, Fourches D, Muratov EN, Oses C, Rasch K, Tropsha A, Curtarolo S (2015) Materials cartography: representing and mining materials space using structural and electronic fingerprints. Chem Mater 27:735–743
23. Schütt K, Glawe H, Brockherde F, Sanna A, Müller K, Gross E (2014) How to represent crystal structures for machine learning: towards fast prediction of electronic properties. Phys Rev B 89:205118
24. Seko A, Hayashi H, Nakayama K, Takahashi A, Tanaka I (2017) Representation of compounds for machine-learning prediction of physical properties. Phys Rev B 95:144110
25. de Jong M, Chen W, Notestine R, Persson K, Ceder G, Jain A, Asta M, Gamst A (2016) A statistical learning framework for materials science: application to elastic moduli of k-nary inorganic polycrystalline compounds. Sci Rep 6:34256
26. Legrain F, Carrete J, van Roekeghem A, Curtarolo S, Mingo N (2017) How chemical composition alone can predict vibrational free energies and entropies of solids. Chem Mater 29:6220–6227
27. Medasani B, Gamst A, Ding H, Chen W, Persson KA, Asta M, Canning A, Haranczyk M (2016) Predicting defect behavior in B2 intermetallics by merging ab initio modeling and machine learning. NPJ Comput Mater 2:1–10
28. Ulissi ZW, Medford AJ, Bligaard T, Nørskov JK (2017) To address surface reaction network complexity using scaling relations machine learning and DFT calculations. Nat Commun 8:1–7
29. Graser J, Kauwe SK, Sparks TD (2018) Machine learning and energy minimization approaches for crystal structure predictions: a review and new horizons. Chem Mater 30:3601–3612
30. Kim E, Huang K, Jegelka S, Olivetti E (2017) Virtual screening of inorganic materials synthesis parameters with deep learning. NPJ Comput Mater 3:1–9
31. Wang C, Shen C, Cui Q, Zhang C, Xu W (2019) Tensile property prediction by feature engineering guided machine learning in reduced activation ferritic/martensitic steels. J Nucl Mater 151823. http://doi.org/10.1016/j.jnucmat.2019.151823
32. Wang C, Zhang C, Zhao J, Yang Z, Liu W (2017) Microstructure evolution and yield strength of CLAM steel in low irradiation condition. Mater Sci Eng A 682:563–568
33. Sasikumar T, Rajendraboopathy S, Usha K, Vasudev E (2008) Artificial neural network prediction of ultimate strength of unidirectional T-300/914 tensile specimens using acoustic emission response. J Nondestruct Eval 27:127–133
34. Santos I, Nieves J, Penya YK, Bringas PG (2009) Machine-learning-based mechanical properties prediction in foundry production. In: ICCAS-SICE 2009, pp 4536–4541
35. Sterjovski Z, Nolan D, Carpenter K, Dunne D, Norrish J (2005) Artificial neural networks for modelling the mechanical properties of steels in various applications. J Mater Process Technol 170:536–544

36. Datta S, Pettersson F, Ganguly S, Saxén H, Chakraborti N (2007) Designing high strength multi-phase steel for improved strength-ductility balance using neural networks and multi-objective genetic algorithms. ISIJ Int 47:1195–1203
37. Saxén H, Pettersson F (2006) Method for the selection of inputs and structure of feedforward neural networks. 30(6–7):1038–1045. http://doi.org/10.1016/j.compchemeng.2006.01.007
38. Li X (2003) In: Fonseca CM, Fleming PJ, Zitzler E, Deb K, Thiele L (eds) 2nd international conference on evolutionary multi-criterion optimization. Lecture notes in computer science, LNCS 2632, p 207
39. Pettersson F, Chakraborti N, Saxén H (2007) A genetic algorithms based multi-objective neural net applied to noisy blast furnace data. 7(1):387–397. http://doi.org/10.1016/j.asoc.2005.09.001
40. Zhang E, Yin M, Karniadakis G (2020) Physics-informed neural networks for nonhomogeneous material identification in elasticity imaging. arXiv preprint arXiv:2009.04525
41. Metzbower E, deLoach J, Lalam S, Bhadeshia H (2001) Neural network analysis of strength and ductility of welding alloys for high strength low alloy shipbuilding steels. Sci Technol Weld Joining 6:116–124
42. Shigemori H, Kano M, Hasebe S (2011) Optimum quality design system for steel products through locally weighted regression model. J Process Control 21:293–301
43. Swaddiwudhipong S, Tho KK, Liu ZS, Hua J, Ooi NSB (2005) Material characterization via least squares support vector machines. Model Simul Mater Sci Eng 13:993–1004
44. Huber N, Tsagrakis I, Tsakmakis Ch (2000) Determination of constitutive properties of think metallic films on substrates by spherical indentation using neural networks. Int J Solids Struct 37:6499–6516
45. Huber N, Nix WD, Gao H (2002) Identification of elastic–plastic material parameters from pyramidal indentation of thin films. Proc R Soc Lond A 458:1593–1620
46. Tho KK, Swaddiwudhipong S, Liu ZS, Hua J (2004) Artificial neural network model for material characterization by indentation. Model Simul Mater Sci Eng 12:1055–1062
47. Abdalla JA, Hawileh R (2011) Modeling and simulation of low-cycle fatigue life of steel reinforcing bars using artificial neural network. J Frankl Inst 348:1393–1403
48. Manson SS (1953) Behavior of materials under conditions of thermal stress. In: Heat transfer symposium. University of Michigan Engineering Research Institute, Ann Arbor, pp 9–75
49. Coffin LF Jr (1954) A study of the effects of cyclic thermal stresses on a ductile metal. Trans Am Soc Mech Eng 76:931–950
50. Koh SK, Stephens RI (1991) Mean stress effects on low cycle fatigue for a high strength steel. Fatigue Fract Eng Mater Struct 14(4):413–428
51. Lee JA, Almond DP, Harris B (1999) The use of neural networks for the prediction of fatigue lives of composite materials. Compos Part A Appl Sci Manuf 30:1159–1169
52. Agrawal A, Deshpande PD, Cecen A, Basavarsu GP, Choudhary AN, Kalidindi SR (2014) Exploration of data science techniques to predict fatigue strength of steel from composition and processing parameters. Integr Mater Manuf Innov 3:90–108
53. Zhang L, Lei J, Zhou Q, Wang Y (2015) Using genetic algorithm to optimize parameters of support vector machine and its application in material fatigue life prediction. Adv Nat Sci 8:21–26
54. Ji DM (2011) Study on the fatigue life of P91 steel creep based on support vector machine. Pressure Vessel 28(15):15–21
55. Verma A, Parashar A, Packirisamy M (2018) Atomistic modeling of graphene/hexagonal boron nitride polymer nanocomposites: a review. Wiley Interdisc Rev Comput Mol Sci 8(3):e1346
56. Verma A, Singh VK, Verma SK, Sharma A (2016) Human hair: a biodegradable composite fiber—a review. Int J Waste Resour 6(206):2
57. Verma A, Singh VK (2019) Mechanical, microstructural and thermal characterization of epoxy-based human hair–reinforced composites. J Test Eval 47(2):1193–1215
58. Verma A, Parashar A (2018) Structural and chemical insights into thermal transport for strained functionalised graphene: a molecular dynamics study. Mater Res Express 5(11):115605

59. Verma A, Negi P, Singh VK (2019) Experimental analysis on carbon residuum transformed epoxy resin: chicken feather fiber hybrid composite. Polym Compos 40(7):2690–2699
60. Verma A, Gaur A, Singh VK (2017) Mechanical properties and microstructure of starch and sisal fiber biocomposite modified with epoxy resin. Mater Perform Charact 6(1):500–520
61. Verma A, Parashar A, Packirisamy M (2019) Effect of grain boundaries on the interfacial behaviour of graphene-polyethylene nanocomposite. Appl Surf Sci 470:1085–1092
62. Verma A, Budiyal L, Sanjay MR, Siengchin S (2019) Processing and characterization analysis of pyrolyzed oil rubber (from waste tires)-epoxy polymer blend composite for lightweight structures and coatings applications. Polym Eng Sci 59(10):2041–2051
63. Verma A, Negi P, Singh VK (2018) Physical and thermal characterization of chicken feather fiber and crumb rubber reformed epoxy resin hybrid composite. Adv Civ Eng Mater 7(1):538–557
64. Verma A, Negi P, Singh VK (2018) Experimental investigation of chicken feather fiber and crumb rubber reformed epoxy resin hybrid composite: mechanical and microstructural characterization. J Mech Behav Mater 27(3–4)
65. Chaurasia A, Verma A, Parashar A, Mulik RS (2019) Experimental and computational studies to analyze the effect of h-BN nanosheets on mechanical behavior of h-BN/polyethylene nanocomposites. J Phys Chem C 123(32):20059–20070
66. Jain N, Verma A, Singh VK (2019) Dynamic mechanical analysis and creep-recovery behaviour of polyvinyl alcohol based cross-linked biocomposite reinforced with basalt fiber. Mater Res Express 6(10):105373
67. Verma A, Joshi K, Gaur A, Singh VK (2018) Starch-jute fiber hybrid biocomposite modified with an epoxy resin coating: fabrication and experimental characterization. J Mech Behav Mater 27(5–6)
68. Verma A, Kumar R, Parashar A (2019) Enhanced thermal transport across a bi-crystalline graphene–polymer interface: an atomistic approach. Phys Chem Chem Phys 21(11):6229–6237
69. Verma A, Singh VK (2016) Experimental investigations on thermal properties of coconut shell particles in DAP solution for use in green composite applications. J Mater Sci Eng 5(3):1000242
70. Verma A, Singh VK, Arif M (2016) Study of flame retardant and mechanical properties of coconut shell particles filled composite. Res Rev J Mater Sci 4(3):1–5
71. Verma A, Parashar A, Jain N, Singh VK, Rangappa SM, Siengchin S (2020) Surface modification techniques for the preparation of different novel biofibers for composites. In: Biofibers and biopolymers for biocomposites. Springer, Cham, pp 1–34
72. Rastogi S, Verma A, Singh VK (2020) Experimental response of nonwoven waste cellulose fabric–reinforced epoxy composites for high toughness and coating applications. Mater Perform Charact 9(1):151–172
73. Bharath KN, Madhu P, Gowda TG, Verma A, Sanjay MR, Siengchin S (2020) A novel approach for development of printed circuit board from biofiber based composites. Polym Compos 41(11):4550–4558
74. Verma A, Jain N, Parashar A, Gaur A, Sanjay MR, Siengchin S (2021) Lifecycle assessment of thermoplastic and thermosetting bamboo composites. In: Bamboo fiber composites. Springer, Singapore, pp 235–246
75. Singh K, Jain N, Verma A, Singh VK, Chauhan S (2020) Functionalized graphite-reinforced cross-linked poly (vinyl alcohol) nanocomposites for vibration isolator application: morphology, mechanical, and thermal assessment. Mater Perform Charact 9(1):215–230
76. Verma A, Parashar A (2018) Molecular dynamics based simulations to study the fracture strength of monolayer graphene oxide. Nanotechnology 29(11):115706
77. Verma A, Parashar A (2017) The effect of STW defects on the mechanical properties and fracture toughness of pristine and hydrogenated graphene. Phys Chem Chem Phys 19(24):16023–16037
78. Verma A, Parashar A, Packirisamy M (2018) Tailoring the failure morphology of 2D bicrystalline graphene oxide. J Appl Phys 124(1):015102

79. Singla V, Verma A, Parashar A (2018) A molecular dynamics based study to estimate the point defects formation energies in graphene containing STW defects. Mater Res Express 6(1):015606
80. Verma A, Parashar A (2018) Molecular dynamics based simulations to study failure morphology of hydroxyl and epoxide functionalised graphene. Comput Mater Sci 143:15–26
81. Verma A, Zhang W, van Duin AC (2021) ReaxFF reactive molecular dynamics simulations to study the interfacial dynamics between defective h-BN nanosheet and water nanodroplets. Phys Chem Chem Phys 23:10822–10834
82. Verma A, Parashar A, Packirisamy M (2019) Role of chemical adatoms in fracture mechanics of graphene nanolayer. Mater Today Proc 11:920–924
83. Verma A, Parashar A (2020) Characterization of 2D nanomaterials for energy storage. In: Recent advances in theoretical, applied, computational and experimental mechanics. Springer, Singapore, pp 221–226
84. Verma A, Jain N, Parashar A, Singh VK, Sanjay MR, Siengchin S (2020) Design and modeling of lightweight polymer composite structures. In: Lightweight polymer composite structures: design and manufacturing techniques, Chap 7. Taylor & Francis Group (CRC Press), Boca Raton, pp 193–224
85. Verma A, Jain N, Parashar A, Singh VK, Sanjay MR, Siengchin S (2020) Lightweight graphene composite materials. In: Lightweight polymer composite structures: design and manufacturing techniques, Chap 1. Taylor & Francis Group (CRC Press), Boca Raton, pp 1–20
86. Verma A, Parashar A, Singh SK, Jain N, Sanjay MR, Siengchin S (2020) Modeling and simulation in polymer coatings. In: Polymer coatings: technologies and applications, Chap 16. Taylor & Francis Group (CRC Press), Boca Raton, pp 309–324
87. Verma A, Jain N, Rastogi S, Dogra V, Sanjay MR, Siengchin S, Mansour R (2020) Mechanism, anti-corrosion protection and components of anti-corrosion polymer coatings. In: Polymer coatings: technologies and applications, Chap 4. Taylor & Francis Group (CRC Press), Boca Raton, pp 53–66
88. Verma A, Jain N, Kalpana, Sanjay MR, Siengchin S, Jawaid M (2020) Natural fibers based bio-phenolic composites. In: Phenolic polymers based composite materials, Chap 10. Springer Nature, Singapore, pp 153–168
89. Bharath KN, Madhu P, Gowda TY, Verma A, Sanjay MR, Siengchin S (2021) Mechanical and chemical properties evaluation of sheep wool fiber-reinforced vinylester and polyester composites. Mater Perform Charact 10(1):99–109
90. Marichelvam MK, Manimaran P, Verma A, Sanjay MR, Siengchin S, Kandakodeeswaran K, Geetha M (2021) A novel palm sheath and sugarcane bagasse fiber based hybrid composites for automotive applications: an experimental approach. Polym Compos 42(1):512–521
91. Chaudhary A, Sharma S, Verma A (2022) Optimization of WEDM process parameters for machining of heat treated ASSAB'88 tool steel using response surface methodology (RSM). Mater Today Proc 50:917–922
92. Chaudhary A, Sharma S, Verma A (2022) WEDM machining of heat treated ASSAB'88 tool steel: a comprehensive experimental analysis. Mater Today Proc 50:946–951
93. Verma A, Singh VK (2016) Experimental characterization of modified epoxy resin assorted with almond shell particles. ESSENCE-Int J Environ Rehabil Conserv 7(1):36–44
94. Verma A, Samant SS (2016) Inspection of hydrodynamic lubrication in infinitely long journal bearing with oscillating journal velocity. J Appl Mech Eng 5(3):1–7
95. Verma A, Parashar A, Jain N, Singh VK, Rangappa SM, Siengchin S (2020) Surface modification techniques for the preparation of different novel biofibers for composites. Biofibers Biopolymers Biocomposites 1–34
96. Bisht N, Verma A, Chauhan S, Singh VK (2021) Effect of functionalized silicon carbide nanoparticles as additive in cross-linked PVA based composites for vibration damping application. J Vinyl Add Tech 27(4):920–932
97. Kataria A, Verma A, Sanjay MR, Siengchin S (2022) Molecular modeling of 2D graphene grain boundaries: mechanical and fracture aspects. Mater Today Proc 52:2404–2408

98. Arpitha GR, Verma A, Sanjay MR, Siengchin S (2021) Preparation and experimental investigation on mechanical and tribological performance of hemp-glass fiber reinforced laminated composites for lightweight applications. Adv Civ Eng Mater 10(1):427–439
99. Deji R, Verma A, Kaur N, Choudhary BC, Sharma RK (2022) Density functional theory study of carbon monoxide adsorption on transition metal doped armchair graphene nanoribbon. Mater Today Proc 54(3):771–776
100. Deji R, Verma A, Choudhary BC, Sharma RK (2022) New insights into NO adsorption on alkali metal and transition metal doped graphene nanoribbon surface: a DFT approach. J Mol Graph Model 111:108109. https://doi.org/10.1016/j.jmgm.2021.108109
101. Deji R, Jyoti R, Verma A, Choudhary BC, Sharma RK (2022) A theoretical study of HCN adsorption and width effect on co-doped armchair graphene nanoribbon. Comput Theor Chem 1209:113592
102. Deji R, Verma A, Kaur N, Choudhary BC, Sharma RK (2022) Adsorption chemistry of co-doped graphene nanoribbon and its derivatives towards carbon based gases for gas sensing applications: quantum DFT investigation. Mater Sci Semicond Process 146:106670
103. Verma A, Jain N, Singh K, Singh VK, Rangappa SM, Siengchin S (2022) PVA-based blends and composites. In: Biodegradable polymers, blends and composites. Woodhead Publishing, UK, pp 309–326
104. Verma A, Jain N, Mishra RR (2022) Applications and drawbacks of epoxy/natural fiber composites. In: Handbook of epoxy/fiber composites. Springer, Singapore, pp 1–15
105. Lila MK, Verma A, Bhurat SS (2022) Impact behaviors of epoxy/synthetic fiber composites. In: Handbook of epoxy/fiber composites. Springer, Singapore, pp 1–18
106. Verma A, Jain N, Sanjay MR, Siengchin S (2022) Viscoelastic properties of completely biodegradable polymer-based composites. In: Vibration and damping behavior of biocomposites. CRC Press, Boca Raton, pp 173–188

Machine Learning Prediction for the Mechanical Properties of Lightweight Composite Materials

Lin Feng Ng and Mohd Yazid Yahya

1 Introduction

The exploitation of the earth has continuously increased at a worrying rate over the past decades. Undoubtedly, the extensive consumption of natural resources will leave a profound negative impact on the global climate and biodiversity. Therefore, researchers and engineers are being surrounded by this issue and urged to take critical steps to avoid environmental degradation. A majority of the polymer matrix on the market are derived from fossil fuels. The radical increase in the production of these polymer matrices eventually further increases the depletion rate of this natural resource. Furthermore, these fossil fuel-based polymer matrices show a high persistence in the environment. The high persistence characteristic of the polymer matrices in the environment could lead to waste accumulation after their service life for a long period. All of these factors contribute to environmental degradation and pollution. On this note, it is important to reduce the use of petroleum-based polymer matrices to protect our mother earth and preserve the environment.

One of the alternative ways to reduce the reliance on petroleum-based polymer matrices is to incorporate fibre reinforcement to form fibre-reinforced polymers (FRPs). The amount of polymer matrices can be reduced by introducing a certain percentage of fibre reinforcement in FRPs. In addition to reducing the use of polymer matrices, the incorporation of fibre reinforcement in FRPs was found to improve the mechanical, physical and thermal properties of the materials. Thus, the addition of

L. F. Ng (✉) · M. Y. Yahya
Centre for Advanced Composite Materials (CACM), Universiti Teknologi Malaysia, 81310 Johor Bahru, Johor, Malaysia
e-mail: linfeng@utm.my

M. Y. Yahya
e-mail: yazidyahya@utm.my

Faculty of Mechanical Engineering, Universiti Teknologi Malaysia, 81310 Johor Bahru, Johor, Malaysia

V. Kushvaha et al. (eds.), *Machine Learning Applied to Composite Materials*,
Composites Science and Technology, https://doi.org/10.1007/978-981-19-6278-3_6

fibre in FRPs is regarded as a promising and appropriate strategy to combat the environmental issues related to waste accumulation and pollution. FRPs are composed of multiple phases which are in intimate contact. Each of the solid phases in FRPs has its own distinctive properties. Since FRPs are formed by combining reinforcement and polymer matrix, it offers the opportunity to the engineers to tailor the mechanical properties of the materials to meet the criteria for certain engineering applications. Nevertheless, the mechanical properties of FRPs are not only governed by fibre and matrix properties. Specifically, fibre-matrix compatibility, fibre orientation, fibre content, and environmental factors also significantly influence the mechanical properties of composites. As of today, intensive research studies have been experimentally performed to optimise the mechanical performance of FRPs [1–5].

Mechanical characterisation of FRPs through experimental works or the prediction using artificial intelligent techniques is of utmost importance in composite design and development. The importance of mechanical characterisation has driven innumerable researchers to conduct investigations on this topic. Generally, the research studies are often dedicated to a particular topic such as the manufacturing process, combination of different constituents in FRPs, process parameters or environmental conditions. In addition, some of the research studies focused on optimising FRPs through chemical or physical modifications of the reinforcement. The findings obtained from these literature studies definitely can provide an insight into the FRPs and serve as valuable information to predict the characteristics of particular FRPs. Nonetheless, many research studies on FRPs were not performed with an intention for certain engineering applications, thus limiting the use of such FRPs in real-life applications.

The experimental investigation has been the conventional way to explore the mechanical performance of FRPs. However, an experimental investigation is time-consuming and often requires a high cost for material preparation and equipment. Besides, introducing a new manufacturing technique or material combination in FRPs involves intensive uncertainties with hefty costs before reaching the goals and objectives. However, it is worth mentioning that the hefty cost can be reduced with the help of fundamental and in-depth knowledge in the field of composite materials. To completely resolve the problem, searching for a low-cost and efficient way to predict the mechanical properties of FRPs is essential. By applying the mathematical and computational models, the best combination of the different phases in FRPs can be determined to provide the optimum mechanical properties to the materials in low-cost and time-saving ways. In this technological era, a more advanced and promising non-destructive testing technique to predict the mechanical performance of FRPs has been well-established. This emerging technology is not time-consuming and does not require high expenses for material preparation and equipment. With the advancement in the field of composite materials and computer science, the use of machine learning to predict the mechanical properties of FRPs has been initiated. By using machine learning to predict the material properties for FRPs, the hefty cost and the time-consuming experimental works can be avoided.

2 Background of FRPs

Composites have been recognised as well-known materials for the past few decades owing to their several promising characteristics compared to metal alloys. The intention of incorporating reinforcement in the FRPs is to reduce plastics, which could cause disastrous consequences to the environment. In the aerospace industry, FRPs are the third generation of aircraft materials to replace conventional metal alloys to enhance the fatigue performance of the aircraft components. Furthermore, due to the lightweight characteristic of FRPs, the energy efficiency of the aircraft can be tremendously increased. The achievement in the aerospace industry has stimulated the idea of employing FRPs in the automotive industry. In recent years, FRPs have found applications in automotive industries such as interior door panels, bumpers, transmission tunnels and roof pillars. The most popular types of reinforcement for FRPs include glass, carbon and aramid fibres. These synthetic fibres have gained wide acceptance mainly because of their high strength and stiffness. In particular, glass fibre is regarded as the successor over carbon and aramid fibres as the cost of glass fibre is relatively lower than carbon and aramid fibres. However, the radical use of synthetic fibres has raised ecological and environmental issues as these synthetic fibres are non-biodegradable and harmful to human health. Thus, the drop in reliance on synthetic fibres is essential to preserve and protect our environment.

To further reduce the cost of the FRPs, the addition of highly available filler in FRPs is considered one of the alternative ways [6]. In addition to cost reduction, the presence of filler can help to improve the overall properties of the FRPs, particularly the wear resistance and stiffness. Examples of filler that are commonly seen include molybdenum disulfide (MoS_2), copper oxide (CuO), copper sulfide (CuS), aluminium oxide (Al_2O_3), graphite, etc. Nowadays, the increasing voice for bio-products has changed the research direction towards green materials. On this note, rice husk particulate has been employed as one of the potential fillers for FRPs due to its high availability. Rice husk is an agricultural waste generated from human food processing. Approximately 600 million tonnes of rice husk are produced each year, accounting for 20 wt% of rice production [7]. With no potential use, rice husk left as agricultural waste has led to disposal problems. Therefore, encapsulating rice husk as filler in FRPs can resolve the disposal problems and help improve the mechanical properties and wear resistance of FRPs. Moreover, the rice husk consists of 32% cellulose, making this filler has high mechanical strength since cellulose is the main constituent providing the mechanical strength to the cellulosic materials.

In recent years, the number of research studies focusing on natural fibres has increased drastically, mainly driven by the desire to reduce the reliance on synthetic fibre based products. The outcomes of the research studies provide a better understanding of the miscellaneous properties of natural fibres and their associated FRPs. By comprehensively exploring the potential of natural fibres, it has been attested that incorporating natural fibres in the FRPs offers several benefits in terms of environmental concerns and mechanical performance. Aside from the commonly cited virtues of natural fibre composites such as lightweight, biodegradability, carbon dioxide neutral and low abrasion to tools, encapsulating natural fibres in FRPs can improve the acoustic damping and specific mechanical properties, which are even greater than those of glass and carbon fibre-reinforced composites. The remarkable acoustic damping characteristic makes natural fibre composites suitable to be employed in automotive industries where noise attenuation is a major concern. However, it is worth mentioning that the mechanical performance of the natural fibre-reinforced composites greatly relies on fibre properties. Growing conditions, harvesting period, and extraction technique are the external factors that could have a significant impact on fibre properties. In terms of the chemical composition, high amount of cellulose may provide greater strength to the natural fibre, whereas the high lignin content may retard the moisture uptake of the fibres. Experimental investigations on the miscellaneous properties of natural fibre-reinforced composites for over a few decades have successfully made natural fibres utilised in large-scale engineering applications. As a result, the latest trend of prioritising the use of natural fibre in uncounted industrial sectors has led to a drop in reliance on conventional synthetic fibres. Although the benefits offered by natural fibres are attractive, it is undeniable that they are not a demerit-free natural resource. Batch-to-batch variation, low mechanical strength, poor compatibility with hydrophobic polymer matrices and high moisture affinity are those well-known shortcomings of natural fibres. The moisture content in the natural fibres eventually leads to the formation of voids in FRPs during the manufacturing process as the moisture may evaporate when heated. In order to manufacture natural fibre-reinforced composites with their exemplary mechanical properties, their shortcomings must be rectified before being used in load-critical engineering applications. The essential precursor to have an optimum composite strength is intimate contact between fibre and matrix and fibre wettability. In this context, surface pre-treatments or the addition of coupling agents are generally performed to elevate the bonding strength between fibre and matrix and reduce the moisture affinity. Alkali and silane treatments are the two popular methods to enhance the compatibility between natural fibres and hydrophobic polymer matrices. Alkali treatment alters the fibre surface structures, exposing a more reactive functional group of the cellulose and strengthening the mechanical anchoring between the fibre and the polymer matrices. After alkali treatment, a certain amount of unnecessary chemical compositions such as hemicellulose, lignin, wax and impurities is eliminated. Instead of modifying the surface structure of the fibre, silane is a coupling agent which could simultaneously react with the fibre and the matrix, forming a siloxane bridge across the interfacial region, thereby enhancing the fibre-matrix interfacial bonding.

In recent years, kenaf, sisal, pineapple leaf, flax, jute, hemp, and coir fibres have been the most commonly explored cellulosic fibres. Abundancy, cost, economic value and mechanical properties are taken into consideration in the material selection for natural fibre-reinforced composites. In fact, natural fibres can be grouped into several categories based on their origins, such as bast, leaf, seed and fruit. The properties of cellulosic fibres from different origins are varied. For instance, the bast fibres generally have a higher mechanical strength than fibres from other origins, whereas leaf fibres mostly have greater ductility than other types of cellulosic fibres.

In comparison with synthetic fibres, the mechanical properties of natural fibres are considered weak, and the moisture uptake deteriorates their mechanical properties to a further extent. The low mechanical properties and high moisture sensitivity of natural fibres have necessitated the hybridisation of natural with synthetic fibres in FRPs. Innumerable research studies have been conducted to explore the mechanical properties of hybrid composites either reinforced with natural/synthetic fibres or natural/natural fibres [8–12]. In order to maximise the hybridisation effect, the fibres in the composite materials should be strain compatible. Hybrid composites can be formed by merging different types of natural fibres or combining natural with synthetic fibres. Compared to non-hybrid natural fibre-reinforced composites, the mechanical properties and the moisture resistance are apparently improved when hybridising natural with synthetic fibres in FRPs. This is primarily due to the remarkable mechanical properties and moisture resistance of synthetic fibres. Ng et al. [13] studied the mechanical properties and water absorption behaviours of pineapple leaf/glass fibre reinforced hybrid composites. The hybridisation has led to a positive impact on the mechanical properties of the materials. Besides, the water absorption and the diffusion coefficient of the hybrid composites were significantly reduced in comparison with non-hybrid pineapple leaf fibre-reinforced composites. Although synthetic fibres can improve the mechanical properties and moisture resistance, the presence of synthetic fibres in the hybrid composites undermines the environmental friendliness of the materials. Apart from the natural/synthetic fibre based hybrid composites, hybrid composites can also be formed by combining multiple types of natural fibre. In this case, the shortcomings of one fibre could be compensated by the advantages of another natural fibre. As mentioned above, the mechanical properties and the affinity to water molecules are governed by the chemical composition of natural fibres. By combining different types of natural fibre, it is anticipated that the balance in mechanical properties and moisture absorption could be attained without compromising the environmental friendliness.

3 Non-destructive Testing for FRPs

Today, many non-destructive methods, either for inspection or material testing purposes, have been successfully developed. The use of these non-destructive methods for FRPs indeed can help to reduce the inspection and material testing costs. Apart from visual inspection and coin-based tap testing, ultrasonic and radiography inspections are being widely employed in industries to identify the internal damages and flaws of FRPs. It is undeniable that the efficiency of material testing and inspection can be greatly enhanced with the aid of computer science. For instance, the accuracy of the damage detection is highly dependent on the data interpretation of the trained human inspectors. However, the data interpretation might be too subjective, and it might vary with different human inspectors. In addition, it is difficult to assure the accuracy of the outcomes from the tap testing since human factors greatly affect the results. At this point, a computer-aided tap test was introduced to improve the accuracy of conventional tap testing. The acoustic response signal and contact time are recorded using an acceleration sensor, and this response signal is utilised to analyse the internal flaws of FRPs. It can be seen that the development of computer-aided damage inspection techniques is continuously increasing, indicating the essential role of computer science in determining the internal flaws of FRPs.

Apart from computer-aided tap testing, digital image correlation (DIC) is also regarded as another powerful tool for damage assessment of FRPs. It is a user-friendly and cost-effective method to monitor the strain and deformation of the materials, which are considered two important properties in engineering and construction areas [14]. Recently, it can be seen that the use of DIC to measure the strain and deformation of composites has apparently increased, mainly due to the rapid growth of computer and data acquisition technologies. By employing DIC, the images of the composite materials are taken using a camera consisting of a charged-coupled device that can convert the electrical signal to digital values. In comparison with the conventional types of equipment used to monitor the strain evolution, such as extensometer and strain gauge, DIC can offer more accurate results as the conventional equipment is often imperfectly attached to the measurement surface, which could have a negative impact on the accuracy of the results.

Unlike the optical imaging technique such as DIC, acoustic-based ultrasonic and radiographic inspections are among the early established non-destructive inspection technique for engineering materials. The most common internal flaw, such as delamination, can be easily detected by X-ray radiography, provided that the orientation of the delamination is not perpendicular to the X-ray beam. With the technological transformation throughout the years, X-ray computed tomography has been developed to analyse the three-dimensional nature of the damaged FRPs. The computing power of X-ray computed tomography allows the users to gain more understanding of the three-dimensional damage morphology of FRPs. Although X-ray radiography and computed tomography are very efficient in determining the delamination, X-ray is hazardous and harmful to human health. Thus, it is not recommended to utilise X-ray radiography in daily maintenance routine. Ultrasonic inspection is another exhaustive technique to detect the internal damages of composite materials. The ultrasonic inspection device comprises an ultrasonic transducer, transmitter and receiver, and the acoustic signals from the device convey the information that indicates the damage mechanism and morphological behaviour of FRPs. The crack location, crack size, damage area and other failure characteristics can be easily detected based on the acoustic response signals. In addition to damage inspection, advancements in computing power, imaging processing and acquisition techniques have led to the development of innumerable emerging computer-aided material testing techniques for FRPs.

4 Overview of Machine Learning

The recent Fourth Industrial Revolution Machine (Industry 4.0) is transforming the world into a new era with a wealth of data that focuses on interconnectivity, automation, artificial intelligence, and real-time data. Inevitably, big data is one of the technologies, that plays a crucial role in leading Industry 4.0 towards success. In the age of big data, advanced technology is needed to analyse and handle these massive amounts of data intelligently. In this context, the knowledge of artificial intelligence and machine learning has become the key to handling and analysing the data. Machine learning can be considered a branch of artificial intelligence that concentrates on pattern matching and the correlation between different groups of datasets. It is commonly referred to as the most popular recent technology in the fourth industrial revolution since it allows systems to learn and improve from experience without having to be explicitly programmed [15]. When compared to the physical model, machine learning models are more efficient in material design and selection because they are able to manage a massive amount of data to identify the best constituents or desired behaviours. In the material design process, a predictive model will be built using the machine learning approach. This technique has been successfully used to predict the microstructures and plastic deformation of metal alloys [16, 17]. Besides, an attempt has been given to explore the prediction model for atomistic simulations, electronic structures and chemical similarity. Typically, machine learning can be categorised into three major groups, unsupervised learning, supervised learning and reinforcement learning. The algorithms are trained in supervised learning using the labelled datasets for data classification and outcome prediction. Artificial neural networks, naïve bayes, linear regression and logistic regression are examples of supervised learning. Conversely, the labelled datasets are clustered and analysed using machine learning algorithms in unsupervised learning. These algorithms can uncover hidden patterns or data groupings without human interaction. For reinforcement learning, it involves training the machine learning model to make a decision to achieve the goals even in an uncertain and complex environment.

Basically, there are three major steps in building a predictive model using the machine learning approach. Figure 1 shows the three major steps in machine learning to predict the mechanical properties of FRPs. During the first stage, all the data relevant to the properties of FRPs are compiled to build a comprehensive database which is then used to build a machine learning model. The data collection is the single most critical step in building the model, as this step is the backbone of machine learning. It should be emphasised that the accuracy of the outcomes is highly dependent on the quality of the input dataset. In other words, the effectiveness of machine learning is governed by the characteristics of the data and the performance of the algorithm that has been chosen [18]. All the data related to composite properties are analysed to find a recurring pattern, and this pattern will be used to build predictive models using algorithms. Subsequently, these predictive models can be applied to predict the mechanical properties of FRPs. It is easy to handle a massive amount of complex data in machine learning, and this data can be used to predict certain properties which

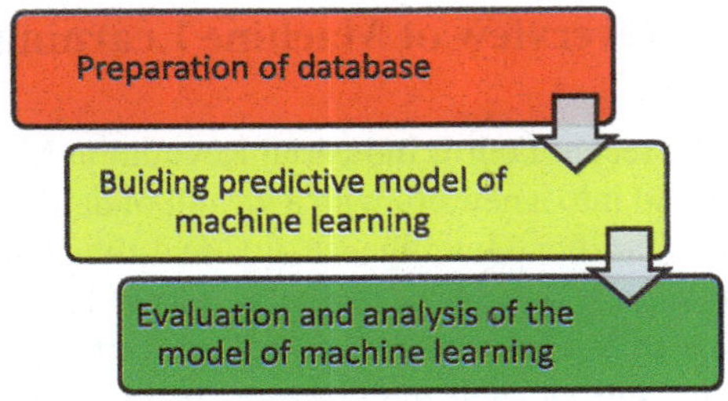

Fig. 1 Three major steps in machine learning for prediction of the properties of FRPs

have not been known yet. Using the historical input data to estimate the unknown properties of composite materials could help to improve the performance and save cost and time. Machines are required to process the data at a high rate and make a decision when the threshold is reached.

In the recent era of big data, we all have experienced the fruitful benefits of artificial intelligence, ranging from medical to engineering. In fact, there are many types of the algorithm in machine learning that are needed to perform the data analysis. Figure 2 summarises the different types of the algorithm in machine learning. Despite a wide variety of machine learning algorithms, selecting the most appropriate learning algorithm for certain applications is challenging. This is because each learning algorithm has its own distinctive functions. The outputs from the different kinds of learning algorithms in a single category may vary, which is highly dependent on the nature of the historical input data. In other words, each learning algorithm has its own strength for certain applications, and it is impossible to apply only a learning algorithm for all situations. Therefore, it is vital to understand the role and applicability of each learning algorithm before applying them in various real-life applications such as cyber security systems, medical diagnosis, internet of things data analysis, smart systems, etc. Understanding the predicting processes in machine learning has become fundamental for obtaining a reliable and accurate outcome. Today, machine learning has been applied for the Covid-19 diagnosis, tracing, screening and prediction, and sometimes it could perform even better than humans [19–21]. Figure 3 displays the different phases in machine learning to predict the future trend of Covid-19 cases. It is anticipated that machine learning in real-world applications will continuously expand in the future.

5 Machine-Learning Prediction of Mechanical Properties of FRPs

With years of endeavour by scientists and engineers, machine learning, an emerging and powerful technology, has risen out of artificial intelligence. It acts as additional support to humans in data analysing to predict processes or properties of materials. Over the years, the rapid growth of data science and machine learning techniques

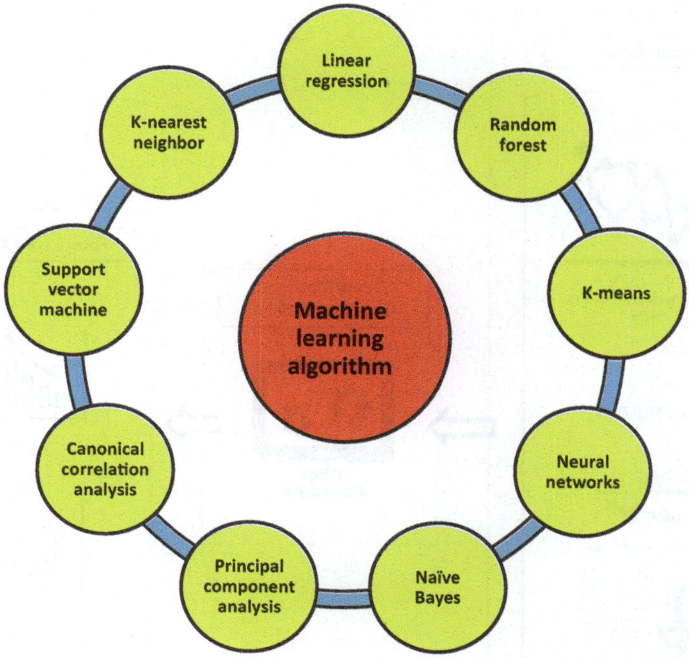

Fig. 2 Different types of algorithm in machine learning

has led to significant growth in the field of composite materials. The prediction of mechanical properties of FRPs using the machine learning approach has been initiated in this technological era. Undoubtedly, the machine learning prediction approach is regarded as an emerging research field for material and computer sciences [23, 24]. In the case of automated fibre placement inspection, the machine learning approach could be more advantageous than the conventional approach. In particular, there are three major advantages of using machine learning for automated fibre placement inspection. First, the machine learning approach can easily identify the boundaries that distinguish different types of defects. The second advantage is that corrective feedback can be obtained by training the machine learning model. Next, the processes in machine learning run in parallel sequence, indicating that the computing time over certain architectures can be prominently enhanced. Figure 4 shows the comparison between the machine learning and traditional modelling approaches for automated fibre placement inspection. Automated fibre placement is the manufacturing process for producing sophisticated advanced aerospace structures that are lightweight and have outstanding performance. This manufacturing process is complicated with several stages of design, planning and manufacturing, and thus the inspection of the material quality is particularly important.

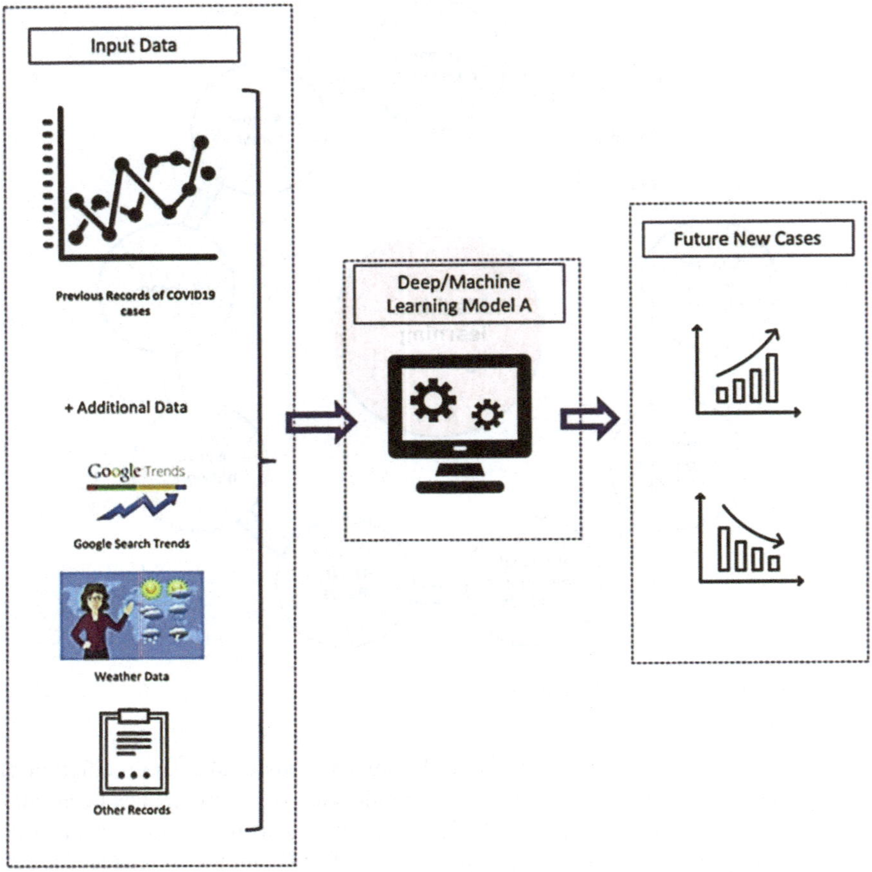

Fig. 3 The different phases of machine learning for the prediction of Covid-19 trend [22]

From the analysis of the previous literature studies, it was noticed that there is an increasing trend of using the artificial intelligence approach to forecast certain material properties. The successful applications of machine learning in various engineering sectors have inspired researchers worldwide to further extend the applications of machine learning. Even though there are a few types of machine learning algorithms, an artificial neural network (ANN), also commonly known as a neural network, has been recognised as an algorithm with remarkable performance when used with complex linear and non-linear relationships. As one of the algorithms in the category of supervised learning, ANN has been shown to be successful in constructing a well-mannered belief model for mechanical properties of materials, mainly due to its capability to capture complicated input and output relationships based on experimental data. Moreover, neural network models are regarded as a less

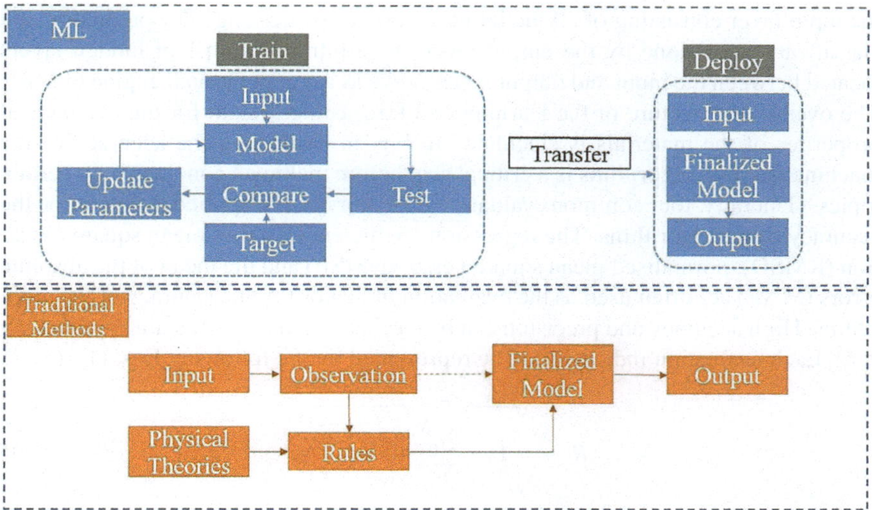

Fig. 4 Comparison between machine learning and traditional modelling approaches (Reprinted with permission obtained from [24])

expensive, less monotonous, more productive, and exceptionally trustworthy solution for estimating material fatigue properties based on data acquired from monotonic testing [25]. When constructing an ANN model, the available data set is separated into two sets, one for training the network and the other for examining the generalisation capabilities of the network.

The idea of developing the mathematical and computational models of ANN is driven by the structural aspects of biological neural networks. This algorithm in machine learning was first introduced in 1943, in which the concept of neurons was developed. The ability to infer a function from data makes ANN models so useful. This is particularly useful when the complexity of the data makes the design of a function difficult to be handled by a human. An ANN can be defined as a mathematical model that consists of a number of highly linked processing components grouped into layers and has been likened to the human brain in terms of architecture and functioning. Figure 5 shows the typical mathematical model of the ANN. Among all the ANN approaches, the multilayer perceptron (MLP) with a back-propagation learning algorithm has attracted the attention of researchers for engineering applications [26]. In comparison with the single-layer perceptron, MLP is more useful for practical applications as it can also be used to learn non-linear functions, whereas single layer perceptron can only be applied for linear functions. With the advancement in the MLP of ANN, researchers have recently paid a lot of attention to the solution of non-linear issues in the physical and mechanical properties of engineering materials. Specifically, MLP consists of three classes: input, hidden, and output layers. It should be noted that MLP contains single input and output layers, but the hidden layers in the predicting stage can be more than one. The input signal will be transmitted to the input layer consisting of all the input factors for processing. The prediction and classification are done by the output layer. An arbitrary number of hidden layers located between the input and output layers serve as a computational engine of MLP. The overall architecture of the learning and forecasting system for the mechanical properties of the materials is elucidated in Fig. 6. Evaluating the accuracy of the machine learning algorithms is a critical step before making a conclusion on certain topics. Generally, four common evaluation indicators can be applied to determine the accuracy of the algorithms. The regression coefficient (R), root mean square deviation (RMSD), normalised mean squared error (NMSE) and the mean of the absolute errors (MAE) are often used as the evaluation indicator for the accuracy of the algorithms. High accuracy and precision can be reached when the errors are approaching zero. Each evaluation indicator can be represented by the following Eqs. (1)–(4):

$$R = \sqrt{1 - \frac{\sum_{i=1}^{N}(x_i - p_i)^2}{\sum_{i=1}^{N}(x_i - \overline{x})^2}} \tag{1}$$

$$RMSD = \sqrt{\frac{\sum_{i=1}^{N}(x_i - p_i)^2}{N}} \tag{2}$$

$$NMSE = \frac{\sum_{d=1}^{D}\sum_{g=1}^{G}[x(d,g) - p(d,g)]^2}{\sum_{d=1}^{D}\sum_{g=1}^{G}[x(d,g)]^2} \tag{3}$$

$$MAE = \frac{\sum_{i=1}^{N}|p_i - x_i|}{N} \tag{4}$$

where x is the true value and p_i is the prediction from the machine learning, N is the total number of data points, $\overline{x}$ is the mean true value.

For the past decades, ANN has been widely employed to predict certain properties of engineering metal alloys. Ozerdem and Kolukisa [27] forecasted the mechanical properties of Cu–Sn–Pb–Zn–Ni cast alloys using the ANN approach. A positive

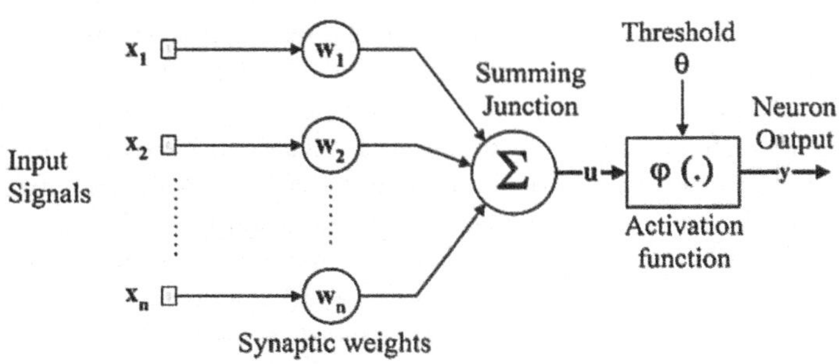

Fig. 5 Mathematical model of artificial neural networks (Reprinted with permission obtained from [27])

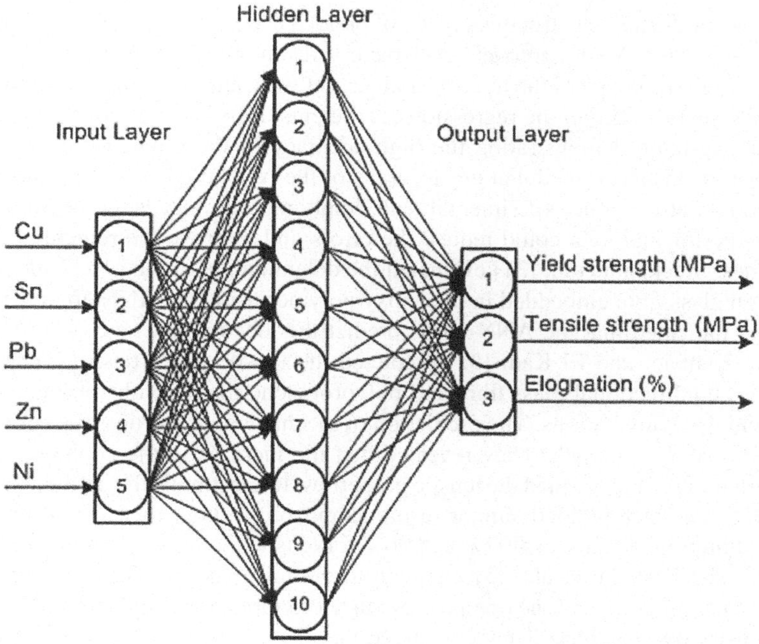

Fig. 6 The overall architecture of the learning and forecasting system (Reprinted with permission obtained from [27])

outcome was obtained in which the ANN successfully predicted the tensile strength, yield strength, and elongation of the cast metal alloys. It was also found that the accuracy of the prediction can be greatly improved by increasing the number of input patterns. Asiltürk and Çunkaş [28] predicted the surface roughness of steel using the ANN and multiple regression approaches. They revealed that the surface roughness value of the steel based on the ANN was more accurate than the multiple regression approach. Hassan et al. [29] forecasted the physical properties of aluminium-copper hybrid metal alloys using the ANN approach. It was concluded that the predicted physical properties of the metal alloys from ANN were in a good correlation with the experimental results with a maximum absolute relative error of approximately 5.99%, indicating that the physical properties of the metal alloys can be accurately predicted using ANN approach without the need of performing the experimental investigation. Thus, the experimental cost and time can be significantly reduced.

Apart from engineering metal alloys, the machine learning approach has also been increasingly employed to predict the mechanical properties of FRPs in order to obtain the best combination of the constituents and various optimum parameters to enhance their mechanical performance. ANN approach is not only limited to metal alloys, it can also be used as a tool to predict miscellaneous properties of FRPs accurately. A recent study performed by Wang et al. [30] focusing on predicting the mechanical properties of the tubular structures made of braided carbon fibre-reinforced composites based on the ANN approach. Satisfactory results were obtained in which the average absolute errors of the training and test set were approximately 5% and 10%, respectively. In addition, the regression coefficients of both the training and test set were above 0.99, demonstrating the high accuracy of the prediction from ANN. However, it was mentioned that the accuracy of the ANN prediction was influenced by the amount of input experimental data, implying that increasing the amount of input experimental data could reduce the errors and improve the reliability of the outcomes. Sreekanth et al. [31] evaluated the delamination of the composite beams based on glass fibre embedded in the epoxy polymer using the ANN approach. The findings indicated that the ANN could predict the location and size of the delamination. Kabbani and El Kadi [32] predicted the tensile properties of composites based on unidirectional glass fibre and polypropylene matrix with varying cooling rates and fibre orientations. The prediction of mechanical properties was performed using the ANN approach. They revealed that the increase in the cooling rate and fibre orientation undermined the tensile properties. Interestingly, the prediction using the ANN approach yielded similar results compared to the experimental findings, with a minimum accuracy of 97% and 90% for tensile modulus and ultimate strain, respectively. Khanam et al. [33] intended to optimise and predict the mechanical and thermal properties of the composites made of graphene nanoplatelets and linear low-density polyethylene. They concluded that the addition of 4 wt% of graphene nanoplatelets and the twin-screw extruder speed at 150 rpm endowed the composites with the highest tensile strength. It was also found that the ANN prediction and experimental findings showed a high correlation coefficient with small error. Rout and Satapathy [34] studied the mechanical and tribological properties of glass fibre-reinforced epoxy composites filled with rice husk filler with varying proportions. The experimental results showed that the addition of rice husk filler improved the hardness, wear resistance, tensile modulus and energy-absorbing capacity, but deteriorated the tensile, flexural and interlaminar shear strength of the composites. The findings predicted from the ANN approach showed an excellent correlation with the experimental results. The error obtained by validating the findings from ANN with experimental results can be reduced by increasing the number of test patterns. An interesting research study has been performed by Al-Assadi et al. [35], focusing on the fatigue properties of composite materials with considering the stress ratio effect using the ANN approach. They revealed very promising results where the fatigue life of the composite materials can be accurately predicted through the ANN approach irrespective of the stress ratio. Fatigue properties of materials are always a challenge in material science, particularly the composite materials, as these materials are composed of two or more different phases, which make the fatigue life assessment becomes complicated and it usually requires a very long time to gain understanding regarding the fatigue properties of the materials. Therefore, there is a great prospect in using the ANN approach to predict the fatigue life of the materials without too much dependence on experimental investigation.

Since natural fibres have gained wide acceptance and popularity among researchers worldwide, the research studies involving natural fibres as the potential reinforcement for the FRPs are continuously growing. With the recent rapid growth of the emerging machine learning in conjunction with Industry 4.0, a growing trend in research studies predicting the mechanical properties of natural fibre-reinforced composites has been noticed. Atuanya et al. [25] explored the mechanical properties of FRPs based on date palm wood fibre-reinforced recycled low-density polyethylene with different fibre contents and sizes using an artificial neural network approach. It was found that the findings predicted by artificial neural networks were in good agreement with the experimental results, showing a coefficient of correlation of above 0.99. Egala et al. [36] predicted the tribological properties of epoxy-based composites reinforced with unidirectional short castor oil fibre using regression and ANN predictive models with single- and multi-hidden layers. They reported that ANN with multiple hidden layers manifested the highest accuracy in predicting the tribological properties, followed by ANN with a single hidden layer and regression model. Keerthi Gowda et al. [37] studied the mechanical properties of untreated and alkali-treated coir/sisal fibre-reinforced polyester composites with varying composite thicknesses and fibre volume fractions. The results demonstrated that the ANN models could accurately predict the mechanical properties of natural fibre-reinforced composites regardless of fibre types, fibre content, composite thickness and chemical treatment, with a correlation coefficient of 0.999 between the predicted properties and the experimental findings. Balaji and Jayabal [38] determined the mechanical properties of zea fibre-based polyester composites with different fibre lengths, fibre contents and moulding pressures. An ANN predictive model was constructed to anticipate the mechanical properties of the composites. They found that the ANN approach can be used to forecast the mechanical properties of the composites accurately with an absolute error percentage of less than 4%. Jayabal et al. [39] identified the influence of fibre length, fibre diameter and calcium carbonate filler content on the mechanical properties of coir fibre-reinforced polyester composites. They revealed that the prediction from the ANN approach was in good agreement with the experimental results with MAE percentages of 0.98%, 0.86% and 0.44% for the tensile, flexural and impact strength, respectively. Ornaghi et al. [40] investigated the dynamic mechanical properties of hybrid composites consisting of sisal/glass fibre embedded in the polyester polymer matrix using experimental and ANN approaches. The hybrid composites were prepared with varying relative fibre ratios. It was revealed that the increase in the glass fibre content augmented the dynamic mechanical performance of the hybrid composites. Overall, the prediction obtained from the ANN approach fitted well with the experimental findings. Shyam et al. [41] mechanical characterised the alkali-treated banyan/glass fibre-reinforced epoxy hybrid composites using ANN and deep neural network approaches. As expected, the incorporation of glass fibre significantly improves the mechanical properties of the hybrid composites. Both approaches have been identified that can accurately predict the mechanical properties of the hybrid composites with an average performance up to 90% in the confusion matrix. Al-Jarrah and AL-Oqla [42] performed a novel research study in predicting the mechanical properties of various natural fibres as a potential reinforcement for

FRPs. Surprisingly, the ANN model successfully predicted the mechanical properties of the natural fibres with an overall percentage for the correct classifications up to 95.6% based on only two properties as the input data: cellulose and moisture content. These findings are indeed very helpful for the designers and scientists to select the most appropriate natural fibres to tailor the mechanical properties of FRPs.

Table 1 summarises the literature studies using the machine learning approach to predict the mechanical properties of FRPs. It can be seen that the predictive model using ANN provides a very promising solution to the issues related to the mechanical properties of FRPs. The optimum parameters and influences of various factors on the mechanical properties of FRPs can be accurately predicted using the ANN approach. Furthermore, with the computational power in the recent technological era, data analysis can be done by using a reliable computer and respective software. Following the findings obtained in the aforementioned literature studies, the prediction of the mechanical properties of FRPs using the machine learning approach is in good agreement with the experimental findings, showing an excellent correlation coefficient. Therefore, it is anticipated that the machine learning approach could substitute the experimental approach to reduce the overall costs and time.

6 Conclusion

FRPs have been well-known materials for the past few decades. Throughout the years, it can be seen that the applications of FRPs are continuously expanding, from the aerospace sector to the automotive and marine sectors. Traditionally, the material properties are investigated and evaluated through experimental works. However, performing an experimental investigation is time-consuming and expensive. In addition, resource and equipment constraints are often the limitations of experimental investigation. These obstacles trigger the researchers to search for other alternative methods to characterise the material properties. Today, one of the most promising mathematical tools that has entered the field of material science is machine learning. In the era of big data, the availability of massive datasets, improvements in algorithms, and exponential increases in computer power have resulted in an unprecedented surge of interest in machine learning. Machine learning methods are now widely used for large-scale classification, regression, clustering, and dimensionality reduction problems involving high-dimensional input data. Indeed, machine learning has demonstrated superhuman powers in a variety of sectors. As a result, machine learning algorithms power many aspects of our daily life. There are several types of learning algorithms in machine learning, and each algorithm has its own benefits and shortcomings for certain practical applications. Therefore, the selection of the most appropriate algorithm has become the critical step prior to the prediction of material properties.

Among the machine learning algorithms, ANN has demonstrated its remarkable performance when used with complex linear and non-linear relationships. ANN is categorised as supervised learning, which is capable of constructing a well-mannered

Table 1 Summary of the literature studies using machine learning approach to predict the mechanical properties of FRPs

Fibre	Matrix	Factor	References
Synthetic fibre			
Carbon	Epoxy	Effect of number of fibre layer	[30]
Glass	Epoxy	Effect of delamination size	[31]
Glass	Polypropylene	Effect of cooling rate and fibre orientation	[32]
Graphene nanoplatelets	Polyethylene	Effect of graphene content and twin screw extruder speed	[33]
Glass	Epoxy	Effect of rice husk filler content	[34]
Glass, graphite, carbon	Epoxy, polyester	Effect of stress ratio and material type	[35]
Natural fibre			
Date palm wood	Polyethylene	Effect of fibre content and size	[25]
Castor oil fibre	Epoxy	Effect of fibre length, nomal load and sliding distance	[36]
Coir, sisal	Polyester	Effect of composite thickness, fibre content and chemical treatment	[37]
Zea	Polyester	Effect of fibre length, fibre content and moulding pressure	[38]
Coir	Polyester	Effect of fibre length, fibre diameter and filler content	[39]
Sisal, glass	Polyester	Effect of relative fibre ratio	[40]
Banyan, glass	Epoxy	Effect of hybridisation	[41]
Flax, pineapple, jute, cotton, sisal, ramie, kenaf	–	Prediction of mechanical properties of various types of natural fibre	[42]

belief model for material properties, primarily due to its capability to capture complicated input and output relationships based on experimental data. Using the ANN to predict the mechanical properties of composite materials will save a lot of money and time for scientists and engineers since they can safely utilise these networks to anticipate material behaviours under different conditions. From the previous literature studies, ANN is not only limited to those conventional metal alloys, but it can be considered an efficient mathematical tool to predict the mechanical, physical and wear properties of the FRPs, displaying a high correlation coefficient with the experimental findings. In contrast, expensive experimental runs take time and can result in mistakes owing to testing imperfections, machinery, or manufacturers. To secure the general growth of manufacturing businesses, encouraging computational modelling to minimise the above-mentioned limitations is essential. Given the success of ANNs in this evaluation, emerging machine learning approaches for process modelling and parameter optimisation of composite material characteristics are strongly recommended.

References

1. Chandrasekar M, Siva I, Kumar TSM et al (2020) Influence of fibre inter-ply orientation on the mechanical and free vibration properties of banana fibre reinforced polyester composite laminates. J Polym Environ 28:2789–2800. https://doi.org/10.1007/s10924-020-01814-8
2. Cordin M, Bechtold T, Pham T (2018) Effect of fibre orientation on the mechanical properties of polypropylene–lyocell composites. Cellulose 25:7197–7210. https://doi.org/10.1007/s10570-018-2079-6
3. Feng NL, Malingam SD, Ping CW, Selamat MZ (2020) Mechanical characterisation of metal-composite laminates based on cellulosic kenaf and pineapple leaf fiber. J Nat Fibers 1–13. http://doi.org/10.1080/15440478.2020.1807437
4. Fiore V, Di Bella G, Valenza A (2015) The effect of alkaline treatment on mechanical properties of kenaf fibers and their epoxy composites. Compos Part B Eng 68:14–21. https://doi.org/10.1016/j.compositesb.2014.08.025
5. Ng LF, Dhar Malingam S, Selamat MZ et al (2020) A comparison study on the mechanical properties of composites based on kenaf and pineapple leaf fibres. Polym Bull 77:1449–1463. https://doi.org/10.1007/s00289-019-02812-0
6. Unal H, Findik F, Mimaroglu A (2003) Mechanical behavior of nylon composites containing talc and kaolin. J Appl Polym Sci 88:1694–1697. https://doi.org/10.1002/app.11927
7. Kim HS, Yang HS, Kim HJ, Park HJ (2004) Thermogravimetric analysis of rice husk flour filled thermoplastic polymer composites. J Therm Anal Calorim 76:395–404. https://doi.org/10.1023/B:JTAN.0000028020.02657.9b
8. Aabdul Khalil HPS, Kang CW, Khairul A et al (2008) The effect of different laminations on mechanical and physical properties of hybrid composites. J Reinf Plast Compos 28:1123–1137. http://doi.org/10.1177/0731684407087755
9. Abd El-baky MA, Attia MA, Abdelhaleem MM, Hassan MA (2020) Mechanical characterisation of hybrid composites based on flax, basalt and glass fibers. J Compos Mater 54:4185–4205. https://doi.org/10.1177/0021998320928509
10. Arpitha GR, Sanjay MR, Senthamaraikannan P et al (2017) Hybridisation effect of sisal/glass/epoxy/filler based woven fabric reinforced composites. Exp Tech 41:577–584. https://doi.org/10.1007/s40799-017-0203-4
11. Feng NL, Malingam SD, Ping CW (2021) Mechanical characterisation of kenaf/PALF reinforced composite-metal laminates: effects of hybridisation and weaving architectures. J Reinf Plast Compos 40:193–205. https://doi.org/10.1177/0731684420956719
12. Feng NL, Malingam SD, Subramaniam K et al (2020) The investigation of the tensile and quasi-static indentation properties of pineapple leaf/kevlar fibre reinforced hybrid composites. Def ST Tech Bull 13:117–129
13. Ng LF, Yahya MY, Muthukumar C (2022) Mechanical characterisation and water absorption behaviors of pineapple leaf/glass fiber-reinforced polypropylene hybrid composites. Polym Compos 43:203–214. https://doi.org/10.1002/pc.26367
14. Wang B, Zhong S, Lee TL et al (2020) Non-destructive testing and evaluation of composite materials/structures: a state-of-the-art review. Adv Mech Eng 12:1–28. https://doi.org/10.1177/1687814020913761
15. Sarker IH, Kayes ASM, Badsha S et al (2020) Cybersecurity data science: an overview from machine learning perspective. J Big Data 7:1–29. https://doi.org/10.1186/s40537-020-00318-5
16. Bessa MA, Bostanabad R, Liu Z et al (2017) A framework for data-driven analysis of materials under uncertainty: countering the curse of dimensionality. Comput Methods Appl Mech Eng 320:633–667. https://doi.org/10.1016/j.cma.2017.03.037

17. Versino D, Tonda A, Bronkhorst CA (2017) Data driven modeling of plastic deformation. Comput Methods Appl Mech Eng 318:981–1004. https://doi.org/10.1016/j.cma.2017.02.016
18. Sarker IH (2021) Machine learning: algorithms, real-world applications and research directions. SN Comput Sci 2:160. https://doi.org/10.1007/s42979-021-00592-x
19. Beck BR, Shin B, Choi Y et al (2020) Predicting commercially available antiviral drugs that may act on the novel coronavirus (SARS-CoV-2) through a drug-target interaction deep learning model. Comput Struct Biotechnol J 18:784–790. https://doi.org/10.1016/j.csbj.2020.03.025
20. Pant S, Singh M, Ravichandiran V et al (2020) Peptide-like and small-molecule inhibitors against Covid-19. J Biomol Struct Dyn 39:2904–2913. https://doi.org/10.1080/07391102.2020.1757510
21. Sipior JC (2020) Considerations for development and use of AI in response to COVID-19. Int J Inf Manage 55:102170. https://doi.org/10.1016/j.ijinfomgt.2020.102170
22. Fard SG, Rahimi HM, Motie P et al (2021) Application of machine learning in the prediction of COVID-19 daily new cases: a scoping review. Heliyon 7:e08143. https://doi.org/10.1016/j.heliyon.2021.e08143
23. Jung KC, Chang SH (2021) Advanced deep learning model-based impact characterisation method for composite laminates. Compos Sci Technol 207:108713. https://doi.org/10.1016/j.compscitech.2021.108713
24. Sacco C, Baz Radwan A, Anderson A et al (2020) Machine learning in composites manufacturing: a case study of automated fiber placement inspection. Compos Struct 250:112514. https://doi.org/10.1016/j.compstruct.2020.112514
25. Atuanya CU, Government MR, Nwobi-Okoye CC, Onukwuli OD (2014) Predicting the mechanical properties of date palm wood fibre-recycled low density polyethylene composite using artificial neural network. Int J Mech Mater Eng 9:7. http://doi.org/10.1186/s40712-014-0007-6
26. Okuyucu H, Kurt A, Arcaklioglu E (2007) Artificial neural network application to the friction stir welding of aluminum plates. Mater Des 28:78–84. https://doi.org/10.1016/j.matdes.2005.06.003
27. Ozerdem MS, Kolukisa S (2009) Artificial neural network approach to predict the mechanical properties of Cu-Sn-Pb-Zn-Ni cast alloys. Mater Des 30:764–769. https://doi.org/10.1016/j.matdes.2008.05.019
28. Asiltürk I, Çunkaş M (2011) Modeling and prediction of surface roughness in turning operations using artificial neural network and multiple regression method. Expert Syst Appl. https://doi.org/10.1016/j.eswa.2010.11.041
29. Hassan AM, Alrashdan A, Hayajneh MT, Mayyas AT (2009) Prediction of density, porosity and hardness in aluminum-copper-based composite materials using artificial neural network. J Mater Process Technol 209:894–899. https://doi.org/10.1016/j.jmatprotec.2008.02.066
30. Wang W, Wang H, Zhou J et al (2021) Machine learning prediction of mechanical properties of braided-textile reinforced tubular structures. Mater Des 212:110181. https://doi.org/10.1016/j.matdes.2021.110181
31. Sreekanth TG, Senthilkumar M, Reddy SM (2021) Vibration-based delamination evaluation in GFRP composite beams using ANN. Polym Polym Compos 29:317–324. https://doi.org/10.1177/0967391121100339
32. Kabbani MS, El Kadi HA (2019) Predicting the effect of cooling rate on the mechanical properties of glass fiber–polypropylene composites using artificial neural networks. J Thermoplast Compos Mater 32:1268–1281. https://doi.org/10.1177/0892705718792351
33. Khanam PN, Almaadeed M, Almaadeed S et al (2016) Optimisation and prediction of mechanical and thermal properties of graphene/LLDPE nanocomposites by using artificial neural networks. Int J Polym Sci 2016:5340252. https://doi.org/10.1155/2016/5340252
34. Rout AK, Satapathy A (2012) Study on mechanical and tribo-performance of rice-husk filled glass-epoxy hybrid composites. Mater Des 41:131–141. https://doi.org/10.1016/j.matdes.2012.05.002
35. Al-Assadi M, Kadi HAE, Deiab IM (2011) Using artificial neural networks to predict the fatigue life of different composite materials including the stress ratio effect. Appl Compos Mater 18:297–309. https://doi.org/10.1007/s10443-010-9158-7

36. Egala R, Jagadeesh GV, Setti SG (2021) Experimental investigation and prediction of tribological behavior of unidirectional short castor oil fiber reinforced epoxy composites. Friction 9:250–272. https://doi.org/10.1007/s40544-019-0332-0
37. Keerthi Gowda BS, Easwara Prasad GL, Velmurugan R (2020) Prediction of mechanical strength attributes of coir/sisal polyester natural composites by ann. J Soft Comput Civ Eng 4:79–105. http://doi.org/10.22115/SCCE.2020.226219.1200
38. Balaji NS, Jayabal S (2016) Artificial neural network modeling of mechanical behaviors of zea fiber-polyester composites. Proc Inst Mech Eng Part E J Process Mech Eng 230:45–55. https://doi.org/10.1177/0954408914539939
39. Jayabal S, Rajamuneeswaran S, Ramprasath R, Balaji NS (2013) Artificial neural network modeling of mechanical properties of calcium carbonate impregnated coir-polyester composites. Trans Indian Inst Met 66:247–255. https://doi.org/10.1007/s12666-013-0255-9
40. Ornaghi HL, Monticeli FM, Neves RM et al (2021) Experimental and artificial neural network approach for prediction of dynamic mechanical behavior of sisal/glass hybrid composites. Polym Polym Compos 29:1033–1043. https://doi.org/10.1177/09673911211037829
41. Shyam S, Kaul S, Kalsara N, Babu TN (2021) Mechanical behaviour and microscopic analysis of epoxy and E-glass reinforced banyan fibre composites with the application of artificial neural network and deep neural network for the automatic prediction of orientation. J Compos Mater. https://doi.org/10.1177/0021998320947136
42. Al-Jarrah R, AL-Oqla FM (2022) A novel integrated BPNN/SNN artificial neural network for predicting the mechanical performance of green fibers for better composite manufacturing. Compos Struct 289:115475. https://doi.org/10.1016/j.compstruct.2022.115475

Ballistic Performance of Bi-layer Graphene: Artificial Neural Network Based Molecular Dynamics Simulations

Kritesh Kumar Gupta, Lintu Roy, and Sudip Dey

1 Introduction

The low strength-to-weight ratio of conventional ballistic proof materials has stimulated the interest of the research community in developing a novel class of materials with high strength and low weight. Graphene, the lightest (density 0.77 mg/m^2) and strongest (tensile strength 130 GPa) material known to date could be a promising candidate for such material advancements. The recent efforts to develop graphene-reinforced materials have revealed that graphene imparts exceptional operational capabilities in the developed composite structures [18, 20, 29, 30]. To this end, in 2014 [19] conducted the laser-induced projectile impact test on the multilayer graphene (thickness $\approx$ 10–100 nm) and observed that multilayered graphene absorbs the projectile kinetic energy 10 times better than the macroscopic steel at the projectile velocity of 600 m/s. Whereas, Ávila et al. [2] and O'Masta et al. [23] investigated the ballistic resistance of graphene reinforced composites. They both reported that the addition of graphene significantly increased the ballistic performance of developed composite materials. The observations drawn from such experimental studies piqued the interest of the research community to explore the ballistic capabilities of graphene and other nano-materials [3, 5, 16, 34]. The recent advent of improved computational prowess has inspired researchers to utilize the molecular dynamics (MD) simulation approach as an alternative to characterize the nano-scale materials experimentally. Over the last few years, the characterization of a wide range of materials through MD simulations of high-velocity ballistic impact is increased in the research community [6–9]. For instance, [36] simulated the high-velocity penetration of silica and nickel

K. K. Gupta (✉) · L. Roy · S. Dey
Department of Mechanical Engineering, National Institute of Technology Silchar, Silchar, Assam, India
e-mail: kritesh_rs@mech.nits.ac.in

S. Dey
e-mail: sudip@mech.nits.ac.in

V. Kushvaha et al. (eds.), *Machine Learning Applied to Composite Materials*,
Composites Science and Technology, https://doi.org/10.1007/978-981-19-6278-3_7

nanoparticles on the pristine and defected graphene sheets. They studied the specific penetration energy of the impacted graphene sheets and discovered that the inherent defects in the graphene sheets significantly reduce graphene's ability to withstand high-velocity impact. Similarly, the follow up atomistic studies of the high-velocity impact on the bilayer graphene and other Van der Waals hetero-structures has revealed that the hybrid graphene structures can absorb the high kinetic energy before failure [24, 26, 33, 35, 38].

The MD simulation has seen widespread use in recent years, and it has been demonstrated to be effective in producing results with acceptable accuracy [11, 28]. However, one of the most noticeable limitations of MD simulation is its computational cost. Even with high-performance computing, the need to perform MD simulations on a large scale to gain a better understanding of material behaviour may be computationally intensive and expensive. This inherent limitation of the conventional MD simulation approach significantly hinders the seamless characterization of nanoscale materials. The combination of machine learning (ML) methods and traditional MD simulation can undoubtedly be effective in addressing this issue. With the rise of machine learning and artificial intelligence in recent years, a large number of studies have been conducted, adapting ML algorithms with MD simulations for a diverse range of materials and reporting the results with high accuracy [12–15, 21, 27].

The concise literature review presented in the preceding paragraphs highlights previous efforts to investigate graphene and its hybrid structures' ballistic performance. However, the studies reported so far only considered the deterministic approach, wherein the graphene structures are tested for a few discrete parametric influences. To have a complete insight into the material behaviour it is essential to gather the responses for large scale unknown sample points within the occurrence bounds of the considered input parameters. Hence, in the present study, we carried out a series of MD simulations of high-velocity impact on the bilayer graphene (BLG) where the impact velocity of a spherical diamond projectile is varied from 1 to 7 km/s (refer to Fig. 1a). The range of impact velocities considered in the present study encompasses the velocity of bullets observed during the military combat and the impact velocity of hails observed during the space exploration and flight [1, 4]. The present study aims to combine the Monte Carlo sampling, MD simulations and artificial neural network (ANN) based machine learning (ML) model to reveal the deep insights of the ballistic performance of BLG (refer to Fig. 1b). In this regard, the impact velocity (V_i) of a projectile is considered as the input parameter, while the corresponding specific penetration energy (E_p^*) of BLG and residual velocity of the projectile after impact (V_r) are considered as the desired quantities of interest for the regression problem. Besides the numerical responses, the post-impact behaviour of BLG is also classified into four different stages viz. R, PP_1, PP_2 and CP, based on the extent of damage to the BLG and the post-impact trajectory of the projectile. The dataset constructed with the MCS based MD simulation is further used to construct the ANN based regression and classification model. The computational capability of ANN based regression model is assessed on the basis of the percentage error between the original and predicted response and for the classification model, true positivity rate is considered as the accuracy measuring criteria. The designed computationally efficient ANN model is deployed to evaluate the performance measures for large scale unknown sample points within the occurrence bounds of the considered impact velocity of projectile (1–7 km/s). The scientific contribution of the present article lies in revealing the deep insights of the ballistic performance of the BLG by successfully integrating the MCS, MD simulation and ANN. That further serves two purposes: first, it proposes a framework to reduce the computational cost associated with large scale MD simulations, and second, it utilises the proposed framework to gather responses for large scale unknown points within the considered range of impact velocity, which would otherwise remain unexplored due to the exorbitant nature of conventional MD simulation. The numerically quantifiable and behavioural findings of this study will pave the way for the development of graphene-based ballistic armour and a new class of composites for high-velocity impact protection.

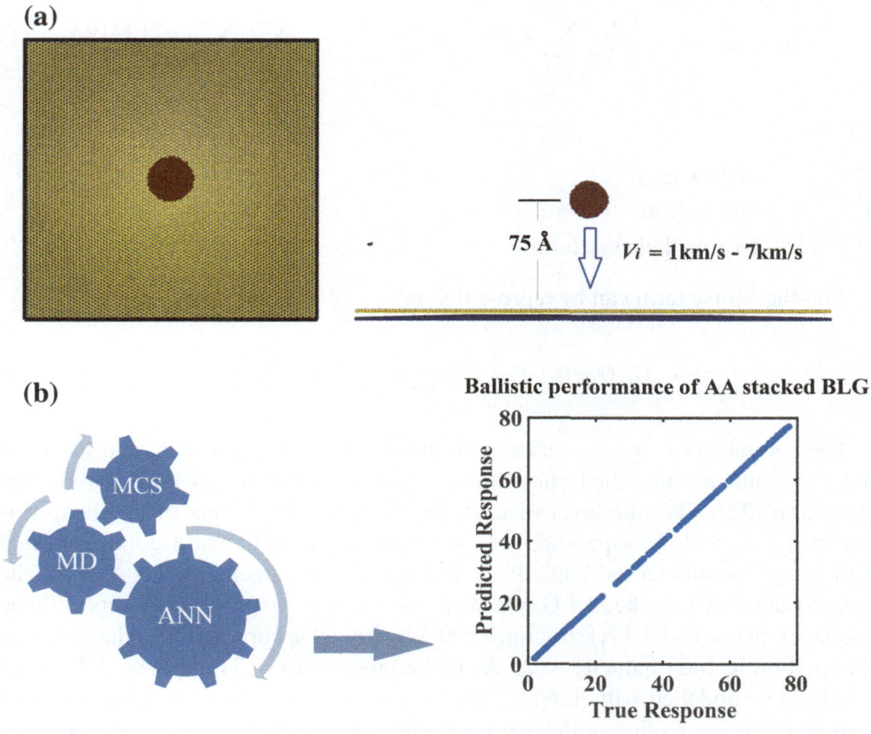

Fig. 1 The machine learning based ballistic behaviour of BLG. **a** The molecular model of BLG (with size = 200 Å × 200 Å) and spherical diamond projectile (with diameter = 25 Å) used in the series of MD simulations, where the impact velocity of the projectile is varied from 1 to 7 km/s. **b** The depiction of the computationally efficient model developed by the successful integration of Monte Carlo sampling, MD simulation, and ANN based machine learning (ML) model

2 Modelling and Simulation

2.1 MD Setup

The inter-atomic and intra-atomic modelling in the MD simulation is performed by using the predefined semi-empirical force fields. The accuracy of results derived from the MD simulation is greatly influenced by the adoption of the appropriate force field. In the present study, we modelled the interatomic interactions of C–C atoms in both layers of BLG by the AIREBO-morse force field. The AIREBO-morse force field consists of the modification of the conventional AIREBO force field, wherein the Lennard–Jones potential is replaced with morse potential to perfectly demonstrate the high-pressure deformation and breaking of bonds. The AIREBO-Morse potential can be mathematically represented as

$$E_{AIREBO-M} = \frac{1}{2}\sum_{i}\sum_{j\neq i} E_{ij}^{REBO} + E_{ij}^{MORSE} + \sum_{k\neq i,j}\sum_{l\neq i,j,k} E_{ij}^{TORSION} \quad (1)$$

where,

E_{ij}^{REBO} REBO term
E_{ij}^{MORSE} Morse term
$E_{ijkl}^{TORSION}$ Torsional energy term.

And the Morse term can be represented as

$$U_{ij}(r) = -\varepsilon_{ij}\left[1 - \left(1 - e^{-\alpha_{ij}(r - r_{ij}^{eq})}\right)^2\right] \quad (2)$$

The parameters ε and r_{eq} denotes minimum energy terms (depth and location) and α denotes the modified curvature of the potential energy with the minimum separation [22]. The interlayer distance between the two layers of graphene was kept at 3.4 Å, and the separation was modelled with the Lennard–Jones potential with energy parameter as 0.00239 eV and distance parameter as 2.5 Å [37]. The square-shaped AA stacked BLG (perfectly overlapped two graphene sheets with the interlayer distance of 3.4 Å) structure (200 Å × 200 Å) is impacted with the spherical diamond projectile (diameter ≈ 25 Å) in the present study. The spherical diamond projectile is modelled with Tersoff [32] force field as it is the best suited potential to model C–C interaction in the diamond. The distance between the top layer of the bilayer structure and the centre of the spherical diamond projectile is kept as 75 Å. The interaction between BLG and diamond is modelled with the morse force field.

The MD simulations are performed in LAMMPS [25] environment and the dumped trajectories are visualized in OVITO [31]. The BLG is modelled through VMD [17]. The MD simulation of high velocity is performed by enforcing the periodic boundary condition in the x and y direction of the BLG while the length of 5 Å on all the four edges of BLG are clamped by setting the force and velocity component to zero. The small iteration time step of 0.5 fs (femtoseconds) is used while performing the MD simulations. At first, the conjugate gradient method is enforced for 25 ps (picoseconds) to minimize the potential energy of the BLG structure. In the next step, the Berendsen thermostat is enforced for the next 25 ps to thermalize the system at room temperature ($\approx$ 300 K). At last, the simulation of high-velocity impact is performed on the BLG structure within an NVE ensemble, wherein the rigid diamond projectile is advanced with high velocity for another 20 ps.

The ballistic performance of the BLG structure is assessed on the basis of the post-impact residual velocity of the projectile (V_r), specific penetration energy (E_p^*) of the BLG, and the behaviour of BLG against the high-velocity impact. The post-impact trajectory of the projectile is observed to determine the residual velocity of the projectile. The negative sign of residual velocity denotes the rebound of the projectile, whereas the positive sign indicates the complete penetration. The specific penetration energy of the BLG is determined by using the following relation [8]

$$E_p^* = \frac{KE_i - KE_r}{M} \tag{3}$$

The KE_i and KE_r refer to the initial and residual kinetic energy of the projectile, respectively. The difference of both kinetic energies is divided by the term M which denotes the mass of the impacted zone. The mass of the impacted zone is determined by the following equation

$$M = \pi.R^2.N_l.\rho_A \tag{4}$$

where R denotes the radius of the impacted zone (same as projectile), N_l denotes the number of layers, and ρ_A denotes the area density of graphene which is taken as $0.77e^{-6}$ kg/m^2 in the present study [16].

2.2 *Artificial Neural Network*

An artificial neural network (ANN) is an adaptive system that learns by utilising interconnected nodes or neurons in a layered structure similar to the human brain. A neural network can learn from data, which enables it to be trained to recognise patterns, classify data, and forecast future events. A neural network abstracts the input by breaking it down into layers. It, like the human brain, can be trained using a large number of examples to recognise patterns in speech or images. Its behaviour is defined by how its elements are connected, as well as the strength, or weights, of those connections. During training, these weights are automatically adjusted according to a predefined learning rule until the artificial neural network correctly performs the desired task. A neural network is a combination of several processing layers that are inspired by biological nervous systems and use simple elements that operate in parallel. It is made up of three layers: an input layer, one or more hidden layers, and an output layer. There are several nodes, or neurons, in each layer, with each layer using the output of the previous layer as its input, so neurons connect the different layers. Each neuron typically has weights that are adjusted during the learning process, and as the weight changes, the strength of that neuron's signal changes [10].

2.2.1 ANN Based Regression

In the present study, the feed-forward back-propagation ANN model is developed for enforcing the successful regression. The model is comprised of two layers, in which 20 hidden neurons is provided in the first layer. The Levenberg–Marquardt training function is employed to train the model. While training the ANN based regression model, the hold-out cross-validation scheme is adopted to prevent the under-fitting or over-fitting of the model. In this regard, the used dataset is divided into three sets, wherein 80% of total samples are used as train data, 10% of total samples are used as test data, and the remaining 10% of the total samples are used as the validation data. The performance of the developed model is assessed on the basis of percentage error observed in the predicted values.

2.2.2 ANN Based Classification

The present study also depicts the utility of ANN based classification in characterizing the behaviour of the considered material system. The ANN based classification model is developed by using the scaled conjugate gradient backpropagation. Similar to the regression model, the classification model is also comprised of two layers and 20 hidden neurons in the first layer, The same hold-out cross-validation scheme is employed for the classification model as well, as employed in the regression model. The performance of the ANN based classification model is assessed on the basis of the confusion matrix, wherein the true positivity rate is observed to justify the applicability of the classification model.

3 Results and Discussion

Prior to performing the MD simulations for a large number of samples, the observations drawn from the high-velocity impact of a diamond projectile on single-layer and bilayer graphene are validated with previous literature findings. The specific penetration energies (E_p^**'s*) of the single-layer graphene and bilayer graphene derived from the MD simulation of high-velocity impact are found in agreement with the corresponding values reported in the past. For instance, [16] reported E_p^* values of 40.8 MJ/kg and 25.2 MJ/kg for single layer and bilayer graphene, respectively, with the high velocity ($V_i = 5$ km/s) impact of rigid fullerene (diameter = 12.1 Å). While, we obtained 42.9 MJ/kg and 23.6 MJ/kg for the single-layer and bilayer graphene, respectively, after replicating the simulations; the obtained values are within 6% error. In another study, [34] reported the penetration energy of single-layer graphene as 4156 eV when a graphene monolayer is impacted ($V_i = 2$ km/s) with a spherical diamond projectile (diameter = 68 Å). With the replication of the simulation, we obtained the penetration energy as 4197.51 eV ($\approx$ 1% error). To ensure the reproducibility of the results obtained from the MD simulations, each simulation is performed five times, and there is no deviation in the results found. With adequate confidence in the findings of the MD simulation, the simulations are performed for the large set of samples. In this regard, the input variable V_i is perturbed within the range of 1–7 km/s (10–70 Å/ps) with the help of the Monte Carlo sampling technique. The sample space with 128 instances is constructed with the help of MCS. In the next step, the MD simulations of high-velocity impact are performed on the AA stacked BLG for each instance in the samples space. The corresponding E_p^* and V_r values are deduced from the findings of the MD simulations as explained in Sect. 2.1. The significant damage to both layers is observed at the impact velocity of 6.1 km/s and the complete penetration of BLG occurred at 6.65 km/s. Previously, complete penetration of the graphene system has been reported at relatively lower impact velocity (1–2 km/s). The increased diameter (140 Å) of the spherical projectile used in the high-velocity impact simulations can explain this [3]. To keep the MD simulations computationally feasible, we used a spherical diamond projectile with a diameter of 25 Å. The deterministic responses concerning the ballistic performance of AA stacked bilayer graphene is presented in Fig. 2, wherein the variation of post-impact residual velocity of the projectile (V_r) and specific penetration energy (E_p^*) of bilayer graphene is represented with respect to the increase in impact velocity (V_i). It can be noticed from Fig. 2a, b that with the increase in V_i the response measures increases. It's worth mentioning that the sign convention for denoting the projectile's residual velocity determines the projectile's post-impact movement direction (in Fig. 2a). The distinct colours of scatter points shown in Fig. 2a reveals the post-impact behaviour of AA stacked bilayer graphene at different impact velocities. The green points denoted as "*R*" corresponds to the rebound of the projectile without any damage to the bilayer graphene sheet, the yellow points denoted as "PP_1" refers to the first stage of partial penetration where the top layer of BLG is penetrated and the projectile rebounds after impact, the orange points denoted as "PP_2" refers to the second stage of partial penetration where both layers of BLG is penetrated and projectile still rebounds after impact, and the red points denoted as "*CP*" refers to the complete penetration of the BLG with no rebound of the projectile. It is evident from Fig. 2 that up to the impact velocity corresponding to the PP_2 stage, the E_p^* values follow a clear increasing trend, whereas after the PP_2 stage E_p^* values exhibits the stagnation in the former increase.

The dataset generated by MCS-driven MD simulations of high-velocity impact on bilayer graphene is used to train, test, and validate the ANN model. To keep over-fitting and under-fitting in check, the hold out cross-validation scheme is used, in which 13 individual samples from the total sample space are separated for testing and validation of the model. The primary screening of the model is performed

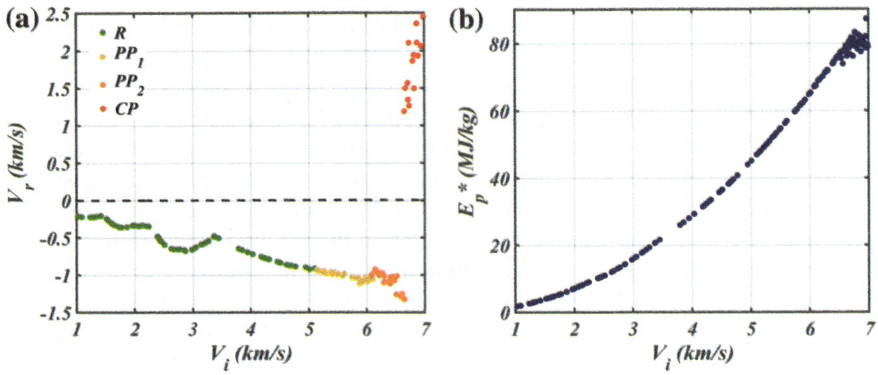

Fig. 2 The deterministic ballistic responses of BLG (derived from 128 MD simulations). **a** The variation in the residual velocity of spherical projectile (V_r) with respect to the increase in impact velocity. **b** The variation in the specific penetration energy (E_p^*) of BLG with respect to the increase in impact velocity

first to determine the applicability of the ANN model in the current scenario. The observations drawn from the primary screening are depicted in Fig. 3.

Figure 3a, b reveals that for both responses, ANN is capable of providing predictions with acceptable accuracy up to an impact velocity of 6.5 km/s. This indicates that the developed model is incapable of predicting the desired quantities associated with the sudden transition of bilayer graphene from PP_2 to CP stage. As the primary screening of the ANN model suggested, we employed the range of impact

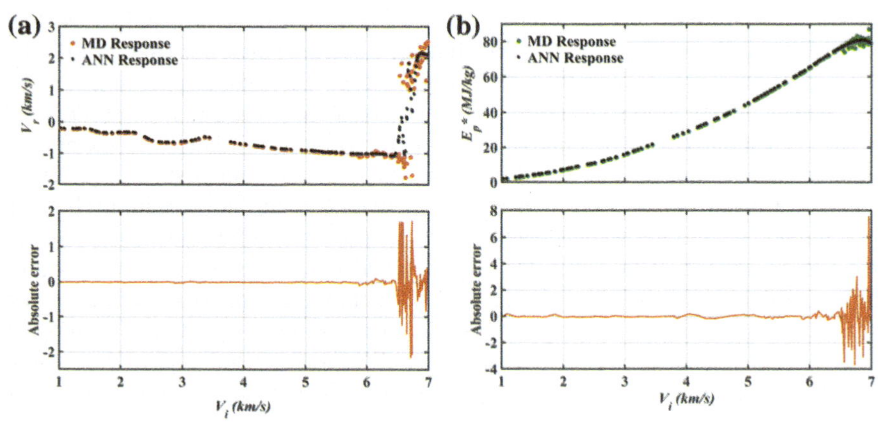

Fig. 3 The primary screening of ANN based regression model. **a** The comparison of the residual velocity of spherical projectile (V_r) drawn from MD simulations and ANN model. **b** The comparison of specific penetration energy (E_p^*) drawn from MD simulations and ANN model. Both the plots revealed that the ANN model is capable of predicting the responses prior to the PP2 stage this can be explained by the sudden transition of BLG from the PP2 stage to the CP stage

velocity from 1 to 6.5 km/s (112 samples) to construct the predictive model. The accuracy of the developed model is assessed on the basis of the percentage error in the predicted quantities. The validation of the developed ANN model is represented in Fig. 4. Figure 4a, b illustrates the model's goodness of fit for each response, whereas Fig. 4c, d displays the created model's generalisation capabilities for each response in terms of % error in the predicted quantities. The validation of the developed ANN model, depicted in Fig. 4, revealed the computational accuracy and capability of the model to generalise the predictions of both the responses of high-velocity impact (with the impact velocity ranging from 1 to 6.5 km/s).

With sufficient confidence in the developed ANN model, the model is further deployed to predict the responses for the large unknown sample points. In this regard, the MCS is used to generate the 10,000 samples with the same range of variation in impact velocity (1–6.5 km/s) as used earlier. The V_r and E_p^* values are determined

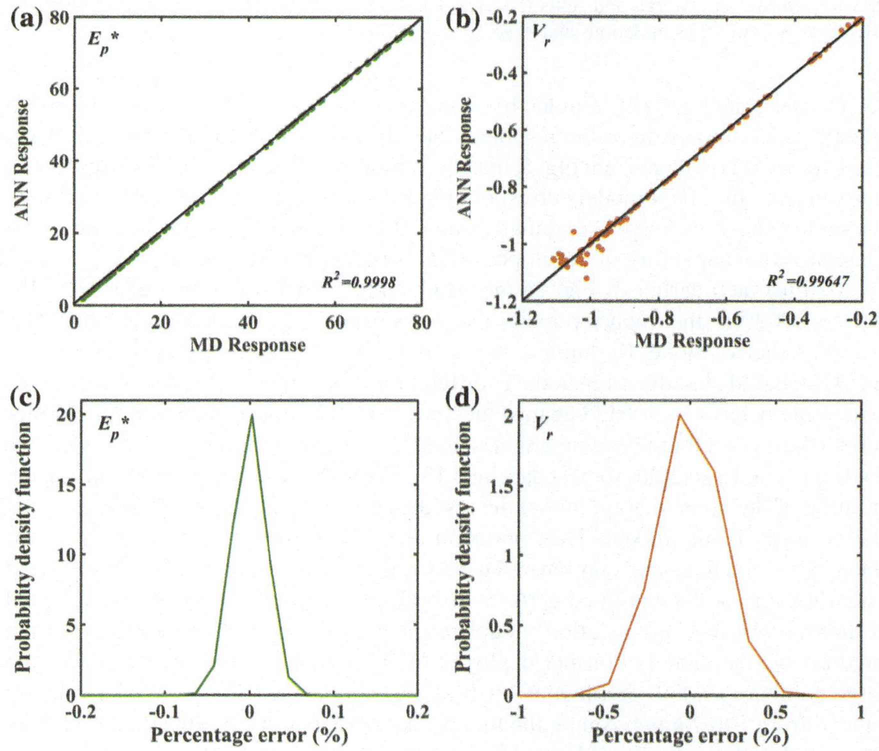

Fig. 4 The validation ANN based regression model developed by considering impact velocities in the range of 1–6.5 km/s. **a** The scatter plot of original E_p^* values and ANN predicted E_p^* values. **b** The scatter plot of original V_r values and ANN predicted V_r values. **c** The probability density function plot for indicating the probability of percentage error in the ANN predicted E_p^* values. **d** The probability density function plot for indicating the probability of percentage error in the ANN predicted V_r values

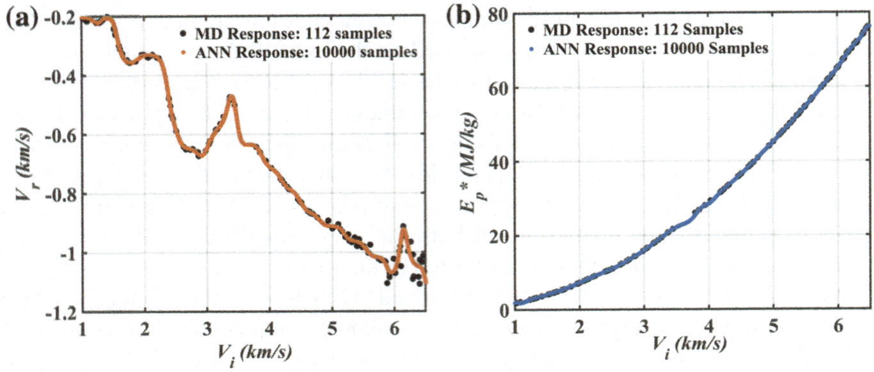

Fig. 5 The ballistic responses predicted for MCS based large scale unknown samples using the developed ANN based regression model. **a** The comparison of the residual velocity of projectile drawn from MD simulations and ANN model. **b** The comparison of specific penetration energy of BLG drawn from MD simulations and ANN model

for the generated 10,000 samples by using the developed ANN models. Figure 5 illustrates a comparison of the responses obtained from MD simulations and ANN predictions. It is evident from Fig. 5, that the variation profile of the ballistic responses derived from the ANN model corresponding to the change in impact velocity follow the same trend as the responses drawn from MD simulations. This indicates the sound generalization capability of developed ANN based regression model.

As illustrated earlier in Fig. 2a the BLG exhibits four different stages (R, PP_1, PP_2, and CP) as the impact velocity increases from 1 to 7 km/s. Besides the ANN based predictive model for numerically quantifiable responses, we also constructed the ANN based classification model to differentiate the impact velocity on the basis of sustaining behaviour of BLG against the high-velocity impact. Similar to the former model the dataset for the classification is divided into three different sets (102 samples for train data, 13 samples for test data, and 13 samples for validation data). During the training of the classification model the test data and validation data is used to assess the accuracy of the model. The confusion matrix corresponding to train data, test data, validation data, and combined datasets is illustrated in Fig. 6a. The ANN based classification model exhibited a 96.1% overall positivity rate in terms of predicted responses. As the classification model resulted in the promising accuracy in the predictions, the model is further deployed to predict the behaviour of BLG for the large number (10,000 samples) of MCS driven samples, where the impact velocity ranged from 1 to 7 km/s. Again, the findings of ANN based classification model are compared with the findings derived from the MD responses (refer to Fig. 6b).

It is evident from Fig. 6b that labels corresponding to the responses gathered from the ANN model perfectly overlap with the labels corresponding to the responses drawn from the MD simulations.

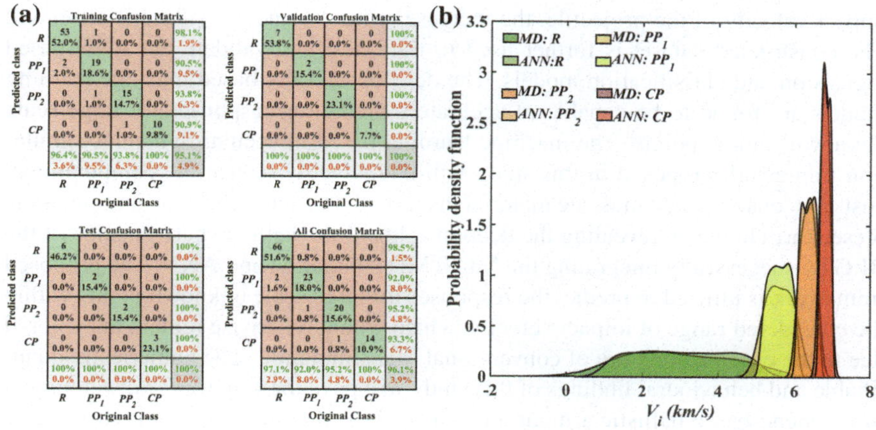

Fig. 6 The validation and outcome of ANN based classification model. **a** The confusion matrices corresponding to the train data, test data, and validation data. The combined confusion matrix is provided with the title "All Confusion Matrix", which emphasizes the generalization capability of the classification model. The model is 96.1% capable of classifying the data with correct labels. **b** The PDF plots comparing the prediction of the classification model (for MCS driven 10,000 unknown samples) and the original responses gathered from series of MD simulations (128 samples)

4 Conclusion and Future Perspective

In the present article, the ballistic performance of AA stacked bilayer graphene is assessed by performing the MD simulations. The series of MD simulations of the high-velocity impact of the spherical diamond projectile (dia = 25 Å) on the AA stacked BLG is performed. However, one of the most noticeable limitations of MD simulation is its computational cost. Even with high-performance computing, the need to perform MD simulations on a large scale to gain a better understanding of material behaviour may be computationally intensive and expensive. To mitigate this lacuna we demonstrated the successful integration of MD simulation with the artificial neural network. The desired responses (post-impact residual velocity of projectile (V_r), specific penetration energy of BLG (E_p^*), and sustaining behaviour of BLG against the high-velocity impact) for the MCS driven sample space are evaluated by performing 128 MD simulations. The impact velocity (V_i) is considered as the control variable in the present study which is perturbed using the MCS approach from 1 to 7 km/s. The deterministic responses drawn from the MD simulations revealed that the complete penetration (*CP* stage) of BLG occurs at the impact velocities beyond 6.5 km/s. Prior to the *CP* stage, the impacted BLG exhibits three different stages viz. *R* stage which denotes the rebound of the projectile without any damage to the BLG, PP_1 stage which denotes the first stage of partial penetration where the top layer of BLG is penetrated and the projectile rebounds after impact, and PP_2 stage which refers to the second stage of partial penetration where both layers of BLG is penetrated and projectile still rebounds after impact. With the increase in the impact velocity of the projectile, the increase in the V_r and E_p^* values is observed. The constructed dataset is further used to train, test and validate the ANN based regression and classification models. The developed ANN-based machine learning models are found to be capable of accurately predicting responses for large scale unknown sample points. The machine learning based molecular dynamics simulation framework proposed in this study will undoubtedly reduce the computational costs associated with nanoscale material discovery. The scientific contribution of the present article lies in revealing the deep insights of the ballistic performance of the BLG by successfully integrating the MCS, MD simulation and ANN. The proposed framework is utilised to predict the responses for large scale unknown points within the considered range of impact velocity, which would otherwise remain unexplored due to the exorbitant nature of conventional MD simulation. The numerically quantifiable and behavioural findings of this study will pave the way for the development of graphene-based ballistic armour and a new class of composites for high-velocity impact protection.

References

1. Abdulhamid H, Bouat D, Colle A, Lafile J, Limido J, Midani I, Papy JM, Puillet C, Spel M, Unfer T, Omaly P (2021) On-ground HVI on a nanosatellite. Impact test, fragments recovery and characterization, impact simulations
2. Ávila AF, Neto AS, Junior HN (2011) Hybrid nanocomposites for mid-range ballistic protection. Int J Impact Eng 38(8–9):669–676
3. Bizao RA, Machado LD, de Sousa JM, Pugno NM, Galvao DS (2018) Scale effects on the ballistic penetration of graphene sheets. Sci Rep 8(1):1–8
4. Børvik T, Olovsson L, Dey S, Langseth M (2011) Normal and oblique impact of small arms bullets on AA6082-T4 aluminium protective plates. Int J Impact Eng 38(7):577–589
5. Dewapriya MA, Meguid SA (2019) Comprehensive molecular dynamics studies of the ballistic resistance of multilayer graphene-polymer composite. Comput Mater Sci 170:109171
6. Dewapriya MA, Miller RE (2021) Energy absorption mechanisms of nanoscopic multilayer structures under ballistic impact loading. Comput Mater Sci 195:110504
7. Dewapriya MA, Miller RE (2020) Molecular dynamics study of the mechanical behaviour of ultrathin polymer–metal multilayers under extreme dynamic conditions. Comput Mater Sci 184:109951
8. Dewapriya MA, Miller RE (2021) Molecular dynamics study of the penetration resistance of multilayer polymer/ceramic nanocomposites under supersonic projectile impacts. Extreme Mech Lett 44:101238
9. Dewapriya MA, Miller RE (2020) Superior dynamic penetration resistance of nanoscale multilayer polymer/metal films. J Appl Mech 87(12):121009
10. Grillakis MG, Koutroulis AG, Alexakis DD, Polykretis C, Daliakopoulos IN (2021) Regionalizing root-zone soil moisture estimates from ESA CCI soil water index using machine learning and information on soil, vegetation, and climate. Water Resour Res 57(5):e2020WR029249
11. Gupta KK, Mukhopadhyay T, Roy A, Dey S (2020) Probing the compound effect of spatially varying intrinsic defects and doping on mechanical properties of hybrid graphene monolayers. J Mater Sci Technol 50:44–58
12. Gupta KK, Mukhopadhyay T, Roy A, Roy L, Dey S (2021) Sparse machine learning assisted deep computational insights on the mechanical properties of graphene with intrinsic defects and doping. J Phys Chem Solids 155:110111

13. Gupta KK, Mukhopadhyay T, Roy L, Dey S (2022) Hybrid machine-learning-assisted quantification of the compound internal and external uncertainties of graphene: towards inclusive analysis and design. Mater Adv 3(2):1160–1181
14. Gupta KK, Roy L, Dey S (2022) Hybrid machine-learning-assisted stochastic nano-indentation behaviour of twisted bilayer graphene. J Phys Chem Solids 167:110711
15. Gupta KK, Roy L, Dey S (2022) Machine learning-based molecular dynamics simulations of monolayered graphene. In: Recent advances in computational and experimental mechanics, vol II. Springer, Singapore, pp 251–263
16. Haque BZ, Chowdhury SC, Gillespie JW Jr (2016) Molecular simulations of stress wave propagation and perforation of graphene sheets under transverse impact. Carbon 102:126–140
17. Humphrey W, Dalke A, Schulten K (1996) VMD: visual molecular dynamics. J Mol Graph 14(1):33–38
18. Kwon SR, Harris J, Zhou T, Loufakis D, Boyd JG, Lutkenhaus JL (2017) Mechanically strong graphene/aramid nanofiber composite electrodes for structural energy and power. ACS Nano 11(7):6682–6690
19. Lee JH, Loya PE, Lou J, Thomas EL (2014) Dynamic mechanical behavior of multilayer graphene via supersonic projectile penetration. Science 346(6213):1092–1096
20. Li Z, He Q, Xu X, Zhao Y, Liu X, Zhou C, Ai D, Xia L, Mai L (2018) A 3D nitrogen-doped graphene/TiN nanowires composite as a strong polysulfide anchor for lithium-sulfur batteries with enhanced rate performance and high areal capacity. Adv Mater 30(45):1804089
21. Mukhopadhyay T, Mahata A, Dey S, Adhikari S (2016) Probabilistic analysis and design of HCP nanowires: an efficient surrogate based molecular dynamics simulation approach. J Mater Sci Technol 32(12):1345–1351
22. O'Connor TC, Andzelm J, Robbins MO (2015) AIREBO-M: a reactive model for hydrocarbons at extreme pressures. J Chem Phys 142(2):024903
23. O'Masta MR, Russell BP, Deshpande VS (2017) An exploration of the ballistic resistance of multilayer graphene polymer composites. Extreme Mech Lett 11:49–58
24. Peng Q, Peng S, Cao Q (2021) Ultrahigh ballistic resistance of twisted bilayer graphene. Curr Comput-Aided Drug Des 11(2):206
25. Plimpton S (1995) Fast parallel algorithms for short-range molecular dynamics. J Comput Phys 117(1):1–9
26. Qiu Y, Zhang Y, Ademiloye AS, Wu Z (2020) Molecular dynamics simulations of single-layer and rotated double-layer graphene sheets under a high velocity impact by fullerene. Comput Mater Sci 182:109798
27. Roy A, Gupta KK, Dey S (2022) Probabilistic investigation of temperature-dependent vibrational behavior of hetero-nanotubes. Appl Nanosci 1–3
28. Roy A, Gupta KK, Naskar S, Mukhopadhyay T, Dey S (2021) Compound influence of topological defects and heteroatomic inclusions on the mechanical properties of SWCNTs. Mater Today Commun 26:102021
29. Shingare KB, Kundalwal SI (2019) Static and dynamic response of graphene nanocomposite plates with flexoelectric effect. Mech Mater 134:69–84
30. Stankovich S, Dikin DA, Dommett GH, Kohlhaas KM, Zimney EJ, Stach EA, Piner RD, Nguyen ST, Ruoff RS (2006) Graphene-based composite materials. Nature 442(7100):282–286
31. Stukowski A (2009) Visualization and analysis of atomistic simulation data with OVITO—the open visualization tool. Modell Simul Mater Sci Eng 18(1):015012
32. Tersoff J (1988) New empirical approach for the structure and energy of covalent systems. Phys Rev B 37(12):6991
33. Tian H, Zhang B, Li QM (2019) Ballistic response of hexagonal boron nitride monolayer under impact of a nano-projectile. Mech Mater 133:1–2
34. Xia K, Zhan H, Hu DA, Gu Y (2016) Failure mechanism of monolayer graphene under hypervelocity impact of spherical projectile. Sci Rep 6(1):1
35. Yang X, Zhang B (2021) Twisted bilayer graphene/h-BN under impact of a nano-projectile. Appl Surf Sci 538:148030

36. Yoon K, Ostadhossein A, Van Duin AC (2016) Atomistic-scale simulations of the chemomechanical behavior of graphene under nanoprojectile impact. Carbon 99:58–64
37. Zhang X, Chen Z, Chen H, Xu L (2021) Comparative studies of thermal conductivity for bilayer graphene with different potential functions in molecular dynamic simulations. Results Phys 22:103894
38. Zheng S, Cao Q, Liu S, Peng Q (2019) Atomic structure and mechanical properties of twisted bilayer graphene. J Compos Sci 3(1):2

Quantifying the Sensitivity of Input Parameters in an ANN-Based Committee Networks Model for Estimation of Steel Girder Bridge Load-Ratings

Fayaz A. Sofi, Irqab Farooq, Javed A. Bhat, and Manzoor A. Tantray

Notations

CN	Committee networks;
D	Dead load effect;
d_e	Barrier-edge distance;
DF	Live load distribution factor;
f_c'	Compressive strength of concrete;
f_y	Yield stress of steel;
K_g	Longitudinal stiffness parameter;
L	Span length; live load effect in Eq. (1)
L_{ANNs}	Total number of ANNs in CN model;
L_{FEM}	FEA result for live load effect;
MSE	Mean square error;
n_b	Number of girders;
RF	Load rating factor;
RF_{ANN}	RF predicted by ANN;
RF_{FEM}	RF obtained from 3D FE analysis;

F. A. Sofi (✉) · I. Farooq · J. A. Bhat · M. A. Tantray
Department of Civil Engineering, National Institute of Technology Srinagar, Hazratbal, Srinagar, Jammu and Kashmir 190006, India
e-mail: sofifayaz@nitsri.ac.in

I. Farooq
e-mail: irqab_2018bciv010@nitsri.ac.in

J. A. Bhat
e-mail: bhat_javed@nitsri.ac.in

M. A. Tantray
e-mail: matantray@nitsri.ac.in

V. Kushvaha et al. (eds.), *Machine Learning Applied to Composite Materials*, Composites Science and Technology, https://doi.org/10.1007/978-981-19-6278-3_8

s Girder spacing;
t_s Deck thickness;
α Skew angle

1 Introduction

In girder-type bridges, the load rating is critical for bridge management and capacity-evaluation decisions, such as load posting, overweight vehicle permits, rehabilitation and replacement prioritizations and needs. The practicing bridge engineering community primarily relies on the load rating for the safe-load carrying assessment of in-service bridges. Accurate and more refined load ratings are increasingly desirable for the older bridge assets designed for lower historical standards, allowing them to remain in service and potentially carry loads even exceeding their original design loads. In the modern bridge specifications [1, 2], the safe-load carrying assessment of girder bridges is based on a simple, component-based loading rating approach that can be rapidly used for the existing bridge inventory. This one-dimensional (1D) approach, commonly known as girder-line analysis, uses transverse load distribution factors to approximate 3D-bridge behavior. In steel girder bridges, each composite girder is treated as an independent free body with simple supports for capacity evaluation. The load distribution factors account for the transverse bridge continuity provided by composite deck-slab and cross-frames. System-level bridge studies [3–10] have demonstrated that AASHTO-based load distribution factors are mostly conservative in various bridge geometries, resulting in conservative load ratings than refined alternative approaches.

Detailed load testing or finite element-based load ratings are more accurate and higher than AASHTO line-girder analysis. The actual 3D bridge behavior involving component interactions leading to system-level complexities that influence load distribution behavior is well captured in these rigorous methods and obscured in line-girder analysis. As per NCHRP Synthesis 453 [11], load rating analysis by refined approaches is employed in a small subset of strategically selected bridges in the inventory. In the survey of 43 DOTs in United States, only 24 used detailed load rating strategies for some selected bridges in their inventories justified as special cases, such as removing load posting restrictions.

In a recent research study [8], artificial neural networks (ANNs)-based models were examined for their potential applications in the load rating predictions of girder-type bridges. The authors proposed ANN-based models for a high prediction accuracy of refined load ratings when used on a subset of steel girder bridges in the Nebraska bridge inventory. ANN-based load rating factors as network outputs were mapped to the girder bridges' influential structural and geometric properties (network inputs) and were referred to as governing parameters. The geometric properties included were girder spacing, span length, presence or absence of cross-frames, longitudinal stiffness, skew angle, barrier distance, number of girders and deck thickness. Among the material properties, the steel-yield strength of girders and compressive strength of the concrete deck was included in the network inputs. Sofi and Steelman [8] proposed and compared the prediction accuracy of two ANNs-based models: (1) committee networks (CN) having combinations of diverse trained ANNs for output averaging and (2) single-best-network consisting of individual best trained ANN. The CN-based approach was superior to the single-best-network model in mitigating prediction errors in load rating predictions on girder bridges.

The present study aims to quantify the sensitivity and significance of the input governing parameters in each of the two ANN-based models proposed by Sofi and Steelman [8], i.e., (1) single-best-network and (2) committee networks (CN). The network inputs to these prediction models were randomly perturbed around their mean values, taking one parameter at a time. The perturbed inputs were generated by adding normally distributed, small random variations to the training data set. Then the variance on the predicted load-rating outputs was used to define the sensitivity of the governing bridge inputs in the prediction models.

2 Methodology

2.1 *Bridge Datasets*

This study obtained a dataset consisting of single span, composite steel girder bridges from the National Bridge Inventory [12], using Nebraska Department of Transportation (NDOT) bridge records to demonstrate the ANN-model (CN and single-best-network models) sensitivity analysis. The selection of bridges was limited to single-span, multi-beam concrete deck composite superstructures carrying up to two lanes of traffic. A dataset of 254 steel girder bridges, 61 existing and 193 hypothetically generated, was randomly selected to provide training and testing datasets for ANN models. The existing set was randomly selected and represented the Nebraska target bridge population sample for the parameters available in NBI [12], i.e., bridge span, width and skew angle. The hypothetical bridge dataset was designed with governing parameters to have parameter distributions similar to the existing set. Table 1 lists the effective ranges of the input governing parameters for ANN models based on the existing bridge subset.

2.2 *Load Rating*

The load rating is used for the assessment of the safe load-carrying capacity of the bridges and is typically expressed as a scale factor of the rating load, referred to as the rating factor (RF > 1 implies a structure has the load-carrying capacity greater than the rating load). The load rating of highway bridges is commonly expressed

Table 1 Governing parameters and their effective ranges

Parameter No.	Bridge parameter	Effective range
1	Span length (L)	20–89 ft (6.10–27.13 m)
2	Girder spacing (s)	32–99 in. (813–2515 mm)
3	Longitudinal stiffness (K_g)	6200–325,800 in.^{4} (2.58 × 10^9–135.61 × 10^9 mm^4)
4	Cross frames	Present or absent
5	Number of girders (n_b)	4–11
6	Skew angle (α)	0–45°
7	Barrier distance (d_e)	(–) 8–34 in. (−203 to 864 mm)
8	Deck thickness (t_s)	5–9 in. (127–228.6 mm)
9	Conc. compressive strength (f_c')	2.5–4 ksi (17.24–27.58 MPa)
10	Steel yield stress (f_y)	30–50 ksi (206.85–344.75 MPa)

at inventory or operating levels [2]. The operating load represents the maximum live load that the bridges can be occasionally subjected to, such as oversized loads requiring special permits. RFs corresponding to operating level has a lower reliability index than the current design practice. The inventory rating level having a reliability index similar to the design specifications represents the lower bound on the safe live load-carrying capacity of the structure that the bridge can safely withstand for an infinite period. The present study quantifies the sensitivity of input governing parameters for moment load ratings at the inventory level.

2.2.1 AASHTO Load Rating

AASHTO bridge specifications [1, 2] provide a simplified one-dimensional (1D) line girder analysis approach for the design and rating of girder bridges. In this simplified 1D method, an individual girder is treated as simply a supported beam, and the transverse continuity of the bridge is taken into account through distribution factors (DFs). The DFs represent the maximum fraction of a single design lane (design load) resisted by each girder. This approach is commonly known as the 'distribution factor method.' AASHTO specifications provide semiempirical equations as functions of bridge geometry (i.e., span length, girder spacing, longitudinal girder stiffness, skew angle and deck thickness) for the calculation of DFs in girder bridges. The girder spacing is the most significant factor influencing the load distribution behavior of girder-type bridges. The AASHTO-based inventory load rating factor (RF; representing the minimum value of the critical girder) can be expressed as a function of girder capacity (C), dead load (D), and live load effect (DF · L), including the impact factor (IM) in Eq. (1).

$$RF = \frac{C - \gamma_d D}{\gamma_L (DF \cdot L + IM)} \tag{1}$$

2.2.2 Refined Load Rating

The precise representation of actual transverse load distribution through plate flexure of the deck slab and consideration of distribution effects by the cross frames etc., can be obtained through 3D finite element (FE) modeling than line-girder-based distribution factors. Thus, when live load effects are directly obtained from a more exact 3D FE analysis, the load rating tends to be less conservative than the AASHTO-based distribution factor approach. The refined load ratings based on 3D FE-based modeling can be obtained by Eq. (2), where notations represent the same quantities as in the AASHTO-based rating, except that live load effects were obtained from 3D FE analyses. Accordingly, there is no need for a distribution factor in Eq. (2), and "$DF \cdot L$" in the former is replaced by L_{FEM} for the live load moments in the latter.

$$RF = \frac{C - \gamma_d D}{\gamma_L (L_{FEM} + IM)} \tag{2}$$

2.3 *FE Modeling and Calibration*

The refined load ratings needed to train ANNs were obtained from detailed FE modeling in ANSYS 19.0 [13]. The schematic flow showing the detailed FE simulation procedure is available elsewhere [9]. All the FE analyses used in the present study for generating refined girder moments were linear elastic. A detailed calibration of the adopted FE modeling methodology was performed against documented results of destructive testing performed on a full-scale laboratory model bridge at the University of Nebraska-Lincoln [14]. The detailed FE model calibration, such as geometry, material simulations, experimental details, etc., for the ultimate validation of laboratory model bridge (referred to here as the Nebraska bridge) are available elsewhere [8, 15]. Figure 1 shows the Nebraska bridge modeling and details, including geometry, FE modeling and calibration against experimental load–deflection results. The geometry creation, selection of elements in numerical models, mapped meshing, the post-processing element results from integration points, etc., were customized using ANSYS parametric design language (APDL), [16]-based macros for automation and consistency, similar to studies [17–21]. The resultant moment for the composite girder was computed within the effective width of the deck as defined in Sect. 4.6.2.6 of Ref. [1]. A summary of FE simulation is provided here.

The concrete deck was simulated with 8-node solids and discretely reinforced by link elements in the top and bottom layers (Fig. 1b). The steel girders and web-stiffener plates were idealized with 4-node shells. The cross-frames were simulated with 2-node Timoshenko beam elements. The composite action between the concrete deck and the steel girders was simulated by multipoint constraint (MPC) rigid beam elements. Simply supported boundary conditions were used on the numerical models

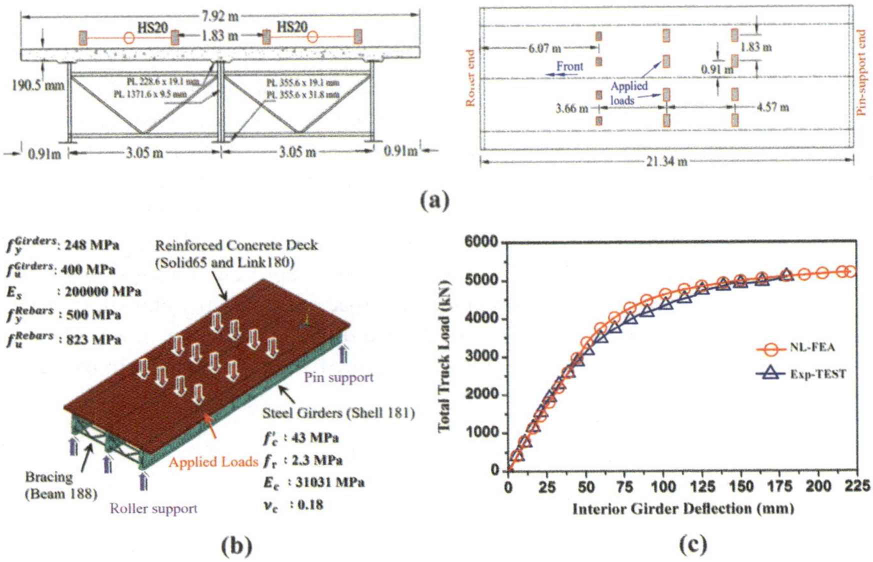

Fig. 1 Nebraska bridge: a bridge geometry and loading; b FE model; c model calibration

by restraining the translational degrees of freedom (i.e., roller and pin support simulations) on the bottom nodes of steel girders at the ends. The material nonlinearities were included only for the calibration model, i.e., the Nebraska bridge using the triaxial failure surface of Willam and Warnke [22] for concrete and isotropic-hardening plasticity for steel with Von Mises yield criteria [23]. For all bridges used for ANN training in this study, linear elastic material properties were assumed (such as for concrete Young's modulus $E_c = 4700\sqrt{f_c'}$ MPa, Poisson's ratio $\upsilon_c = 0.2$ and in steel $E_s = 200{,}000$ MPa, $\upsilon_s = 0.3$) were assumed from [1]. The compressive strength of the concrete deck (f_c') and yield stress of steel (f_y) were based on the typical values documented for the existing bridges. For bridges with unknown information (particularly, old and poorly documented assets) material properties were based on recommendations provided in [2], Sects. 6A5.2.1 and 6A.6.2.1.

The maximum moment responses at the critical sections near the midspan were obtained for standard truck loading (HS20-44). The location of critical sections was based on the influence line analysis. The number of design lanes for each bridge analysis was determined as per Sect. 3.6.1.1 of [1]. In some bridges with roadway widths ranging from 5.49 to 7.32 m (18–24 ft) only two traffic lanes were assumed (refer to Sect. 6A.2.3.2 of [2]). The FE analyses were performed for various combinations of HS20 trucks, transversely positioned in single up to a maximum number of design lanes possible on the bridge. The critical moment values were obtained for each exterior and interior girders. The wheel loads on the FE model for rear and front wheels were simulated using a series of point loads distributed over 500 × 200 mm and 250 × 100 mm patch areas, respectively. As per AASHTO specifications

[1], multiple presence factors were applied to FE-based moment values. The refined load ratings were based on the most critical moments obtained for each exterior and interior girders.

2.4 Committee Neural Networks (CN)

Artificial neural networks (ANNs) mimic the functional abstraction of the biological neurons as robust and adaptive computational units that can approximate multidimensional, complex and nonlinear functions without prior knowledge of the underlying nature or assumptions of the function. When trained with sample cases of inputs and outputs (i.e., patterns), ANN can generalize the relationship of the function by mapping inputs with outputs. Thus, ANNs can predict outputs on similar other inputs that were not encountered during training. Multi-layered feedforward architecture is commonly used for regression analysis in engineering applications [8, 21, 24–27].

A combination of ANNs can be more robust in prediction accuracy than the individual best of the networks that went into the combination [28, 29]. This diverse combination of multiple base-learner networks is referred to as this study's committee network (CN) model. A committee consisting of diverse networks in size and characteristics (such as network architectures, training by different algorithms, variation in data subsets, etc.) can complement generalization ability in learning and provide a more reliable and robust prediction model. Figure 2 shows the CN prediction model for load rating of steel girder bridges proposed by Sofi and Steelman [8]. This CN model consisted of $L_{ANNs} = 4$ separate subcommittee members (each formed by multiple independently trained ANNs). Sofi and Steelman formed the subcommittees by variations of network architecture (one vs. two hidden layers) and learning algorithms (Levenberg–Marquardt, *LM* [30] vs. Bayesian-regularization, BR [31]). For example, a subcommittee designated *I-(m)-O-(LM-k-ANNs)* in Fig. 2 indicates a combination of *k* ANNs having *I* inputs and *O* outputs in each network. Each network has an architecture consisting of *m* neurons in a single hidden layer and is trained by Levenberg–Marquardt's (*LM*) error minimization algorithm. Sofi and Steelman used equal weight factors for all four subcommittees' outputs in a linear combination to obtain CN model prediction (y), as given by Eq. (3):

$$y = \frac{1}{L_{ANNs}} \sum_{j=1}^{L_{ANNs}} y_j \tag{3}$$

where y_j = output from jth individual network in a subcommittee, and an output from the jth individual subcommittee in the CN.

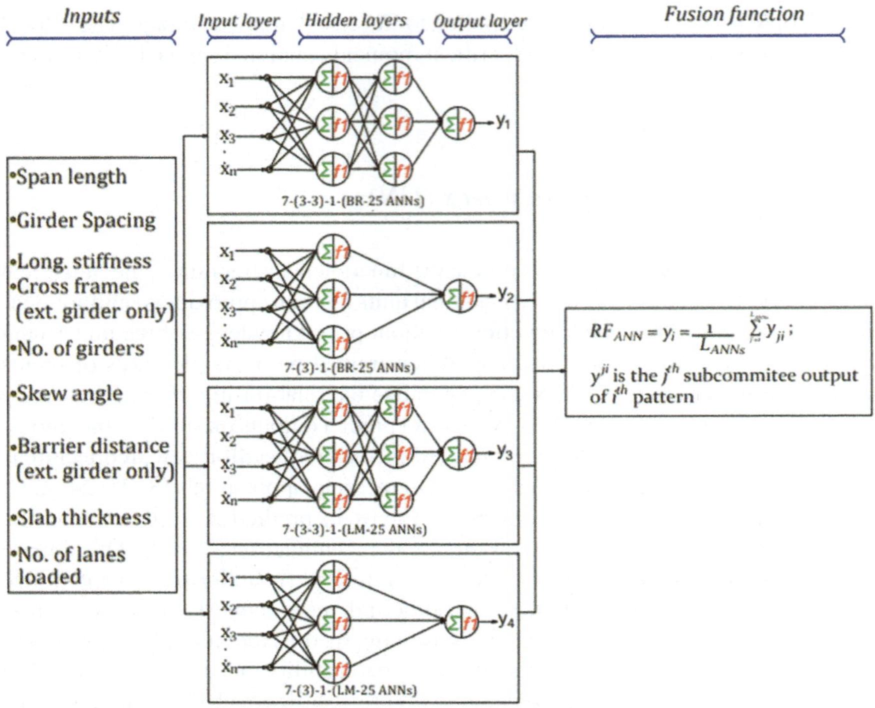

Fig. 2 Committee networks (CN): *I-(m)-O-(LM/BR-k ANNs)* subcommittee consisting of *k* ANNs; I = network inputs; m = hidden layer neurons; O = network outputs; *LM/BR* = Levenberg–Marquardt/Bayesian regularization training algorithms

2.5 Sensitivity Analysis of Network Inputs

Sensitivity analysis is essential for inputs feeding into the network, mainly when limited datasets are available for training ANNs. The requirement for training data patterns increases many folds with more network inputs, leading to a complex input–output mapping having additional network parameters (weights and biases) for new connections within the ANN. Sensitivity analysis of inputs can help minimize the number of inputs without affecting network performance by providing the relative significance of each governing parameter in the input–output mapping relationship of the prediction models. The relative importance of each governing parameter (i.e., ANN inputs) is quantified based on the small random changes in the given input parameters one at a time in the training dataset and measuring its corresponding influence on the network output.

This study's sensitivity analysis of the network inputs was based on the approach provided in [29]. For both CN and single-best-network models, each input parameter was randomly perturbed around their mean values by adding noisy random numbers with normal Gaussian distribution while keeping all other parameters at their mean values. The added input-perturbation had a zero mean, while the variance ($\sigma_{x_i}^2$) for each network input was kept at a constant 10% of the actual variance calculated in the design set data. Accordingly, the change in the network output (i.e., load rating in the present study) for the given input variance ($\sigma_{x_i}^2$) was taken as the sensitivity (S_{x_i}) of the governing parameter x_i, and can be calculated by Eq. 4 [29]:

$$S_{x_i} = \frac{\sum_{p=1}^{P} \sum_{j=1}^{O} \left(y_{jp} - \overline{y}_{jp}\right)^2}{\sigma_{x_i}^2} \tag{4}$$

where, y_{jp} = is the *jth* network output for *pth* pattern in the ANN model. The number of network outputs (O) represented the single moment load rating in the present study. The total number of data patterns (P) and variance ($\sigma_{x_i}^2$) of the input perturbation were used from the design dataset.

3 Discussion of Results

Sofi and Steelman [8] optimized the network architectures for subcommittee members forming a committee networks (CN) model. Figure 2 shows the architectures for each optimized subcommittee of the CN model. Similarly, the single-best-network (i.e., 10-(3)-1-BR) was established to compare sensitivity analysis results on a common design dataset of 100 bridges provided in Appendix along with load ratings. For sensitivity analysis, each of the 10-governing parameters (network inputs) was assigned their average values from the design set. Then a random input perturbation was added to each parameter such that this perturbation had a Gaussian distribution (i.e., mean = 0; variance $\sigma_{x_i}^2$ = 10% of variance in design set for the selected parameter), while keeping all others parameters at their mean values. The resulting load-rating output variance per unit variance of the typical input-network parameter was taken as sensitivity (S_{x_i}) of the governing parameter x_i, calculated using Eq. (4).

Figure 3 shows the results of the sensitivity analysis for (1) committee networks (CN) and (2) single-best-network models. The sensitivity was demonstrated in terms of parameter importance for each of the 10-input governing parameters. The parameter-importance represents relative sensitivity (in terms of percentage) by normalizing sensitivity values with the sum of the sensitivities of all the network inputs. It is evident that girder spacing, barrier-edge distance and bridge span length are the most influencing parameters for the network output (i.e., moment load ratings) for both ANN-based prediction models (Fig. 3). The spacing between girders has the highest importance of 46% in the CN model than the 42% in the single-best-network model.

The barrier edge-distance depicted the highest importance at 46% in the single-best-network model compared to 35% in the CN model. The significance of girder spacing for moment load-rating as the single most influencing parameter can be seen

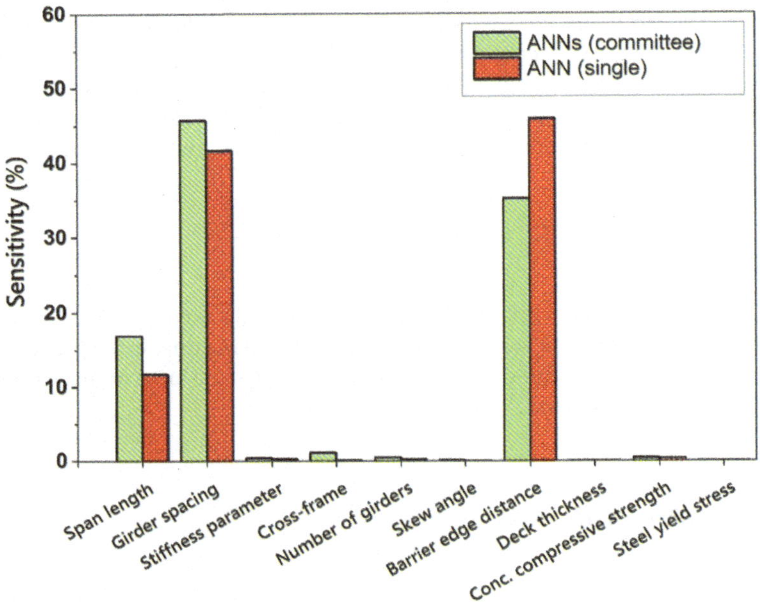

Fig. 3 Relative importance of the governing parameters for moment load-rating prediction using single-best-network and committee networks (CN) models

from the context that it is a well-established analogy for equal importance in the transverse load distribution of girder type bridges [32]. In comparison, the barrier-edge distance can become critical for those girder bridges where the exterior girder controls the load rating, which happens to be the case for 81% of the design set bridges in the present study.

4 Conclusions

This study quantified the sensitivity and demonstrated the importance of the governing parameters (i.e., network inputs) in the moment-based load rating of composite steel girder bridges. The load rating factors were predicted using two ANN-based approaches originally proposed by Sofi and Steelman [8], i.e., (1) single-best-network and (2) committee networks (CN) models. A 10% variance as a random perturbation on a design set of 100 bridges for each network input around the mean input values illustrated the variance in load rating (network output), which was used to define the importance of each governing parameter in the network. Among studied network-input parameters, i.e., girder spacing, span length, presence or absence of cross-frames, longitudinal stiffness, skew angle, barrier distance, number of girders, deck thickness, steel-yield strength and compressive strength of the concrete, the spacing between girders in the CN model and the barrier-edge distance in the single-best-network model were seen as the most influential parameters for moment-based load rating with a sensitivity of about 46% in this study.

Appendix

The representative design set (100 patterns) used for ANN model sensitivity

Bridge number	Type [Existing (E)/Hypothetical (H)]	Design load	Bridge model information (inputs for ANN)					
			L (m)	s (m)	K_g (1 × 10^9 mm^4)	CF (1 or 0)	n_b (No.)	α (deg.)
1	H	H15	11.28	1.68	8.74	1	6	0
2	E	HS20	7.28	0.81	8.58	1	10	0
3	H	H15	10.06	1.57	9.01	1	6	0
4	H	H15	7.01	1.32	4.66	1	5	0
5	H	H15	19.20	2.16	49.21	1	5	0
6	H	H15	14.02	1.07	12.98	1	7	0
7	E	HS20	8.53	1.22	26.67	0	7	0
8	H	H20	24.99	2.06	116.74	1	5	0
9	H	H15	13.72	2.49	24.25	1	4	0
10	E	HS20	8.15	1.19	17.56	0	7	0
11	E	HS20	10.36	1.57	35.51	1	6	30
12	H	HS20	9.75	1.40	25.41	0	6	40
13	H	H20	8.23	0.86	7.25	0	7	0
14	E	HS20	8.79	1.82	30.95	1	5	20
15	H	H20	19.81	1.73	51.16	1	6	45
16	E	Unknown	22.60	2.51	132.45	1	4	0
17	E	HS20	11.89	1.75	40.64	1	6	15
18	E	Unknown	15.24	1.50	36.48	1	7	0
19	H	H15	6.10	1.65	4.13	0	6	35
20	H	HS20	23.16	1.73	83.78	1	6	0
21	E	HS20	8.81	0.86	8.26	1	11	0
22	E	Unknown	6.91	1.79	5.31	0	5	0
23	H	HS20	8.53	2.11	13.20	0	5	0
24	H	HS20	11.58	1.24	23.37	0	6	0
25	H	HS20	23.47	2.08	102.18	1	5	20

(continued)

(continued)

Bridge number	Type [Existing (E)/Hypothetical (H)]	Design load	Bridge model information (inputs for ANN)					
			L (m)	s (m)	K_g (1 × 10^9 mm^4)	CF (1 or 0)	n_b (No.)	α (deg.)
26	H	HS20	7.92	2.16	15.24	0	5	0
27	E	Unknown	19.51	1.68	72.96	1	6	0
28	H	HL93	6.10	1.42	6.44	1	7	0
29	E	HS20	7.32	1.19	11.32	1	7	0
30	H	H15	21.34	1.52	55.17	1	6	0
31	E	HS20	14.84	1.77	58.44	1	6	0
32	H	H20	27.13	1.35	129.71	1	6	0
33	H	HS20	6.40	1.30	5.45	0	7	0
34	H	H15	13.72	1.37	23.68	1	5	0
35	H	HS20	25.60	1.30	131.57	1	6	0
36	E	HS20	11.86	1.81	36.55	1	6	0
37	H	HS20	17.98	1.83	73.79	1	6	15
38	E	HS20	10.62	2.39	21.56	1	5	0
39	H	H15	24.69	1.73	81.06	1	6	0
40	H	H15	6.71	1.30	2.59	1	6	0
41	H	H20	11.58	1.42	14.53	1	7	0
42	H	HS20	22.86	2.13	135.63	1	5	0
43	E	HS20	13.50	2.11	69.18	1	5	0
44	E	Unknown	14.56	1.41	32.76	1	7	20
45	H	H15	8.84	1.42	6.44	1	7	0
46	H	HL93	6.71	1.24	9.28	1	7	0
47	H	H15	24.08	1.88	76.87	1	6	0
48	E	HS20	17.86	1.93	64.85	1	5	0
49	E	HS20	10.67	1.78	43.26	1	6	30
50	E	HS20	17.89	1.79	96.93	1	5	0
51	E	HS20	10.67	2.12	27.52	1	5	30
52	H	H15	8.53	1.55	6.22	0	6	0
53	H	HS20	10.36	1.09	6.82	1	7	0
54	H	H20	19.20	0.89	39.80	1	7	0
55	H	H15	12.19	0.81	9.56	1	10	0
56	E	HS20	7.62	1.79	7.19	1	6	0
57	H	HS20	15.24	1.75	49.74	1	6	10
58	H	HS20	11.58	1.09	12.52	1	7	0
59	H	H20	16.15	1.35	55.17	0	7	0
60	H	HS20	8.23	2.26	20.10	1	4	10
61	H	HS20	12.19	1.12	21.05	0	6	0
62	H	HS20	8.84	1.65	11.95	0	6	0

(continued)

(continued)

Bridge number	Type [Existing (E)/Hypothetical (H)]	Design load	Bridge model information (inputs for ANN)					
			L (m)	s (m)	K_g (1 × 10^9 mm^4)	CF (1 or 0)	n_b (No.)	α (deg.)
63	H	HS20	11.28	1.65	21.06	1	6	0
64	H	H20	11.89	1.42	16.48	1	7	0
65	H	H20	14.33	0.81	24.83	1	11	0
66	E	HS20	16.15	1.22	75.75	1	7	0
67	H	H20	14.33	0.97	19.92	1	7	0
68	E	Unknown	6.25	1.52	5.33	1	6	0
69	H	H20	12.80	0.81	23.68	1	11	0
70	E	HS20	11.86	1.69	43.26	1	6	0
71	E	HS20	18.03	1.54	93.07	1	6	0
72	H	HS20	10.97	1.32	23.92	1	6	0
73	E	HS20	8.53	1.22	18.91	0	7	0
74	H	H15	8.23	1.45	10.63	0	6	0
75	H	HS20	17.07	1.45	49.74	1	6	0
76	H	H15	9.45	0.91	8.74	1	6	15
77	H	HS20	7.62	1.19	7.77	0	6	0
78	E	Unknown	8.84	1.96	25.75	0	5	45
79	H	H15	17.68	2.18	46.55	1	5	0
80	E	Unknown	15.06	1.70	26.29	0	6	0
81	H	H20	7.92	1.35	7.44	1	5	0
82	H	HS20	8.84	1.17	12.28	0	7	0
83	H	H15	6.71	1.75	3.39	1	6	25
84	H	HS20	12.80	2.11	38.12	1	5	0
85	E	HS20	17.98	1.75	90.87	1	6	0
86	H	HS20	11.28	1.91	40.61	0	5	0
87	H	H15	8.23	1.50	7.68	1	6	0
88	H	HS20	9.45	1.50	12.73	0	7	0
89	H	H20	6.71	0.81	8.85	0	9	0
90	H	H15	14.33	2.24	36.55	1	4	0
91	H	H20	8.23	1.60	8.74	1	5	0
92	H	H15	8.53	0.81	6.62	1	7	45
93	E	HS20	7.62	2.29	19.32	1	4	0
94	H	H15	17.37	1.88	49.74	1	6	30
95	H	HS20	17.68	1.68	56.24	1	6	0
96	H	HS20	11.28	0.89	21.56	1	6	25
97	H	HS20	10.97	1.40	18.99	1	7	0

(continued)

(continued)

Bridge number	Type [Existing (E)/Hypothetical (H)]	Design load	Bridge model information (inputs for ANN)					
			L (m)	s (m)	K_g (1 × 10^9 mm^4)	CF (1 or 0)	n_b (No.)	α (deg.)
98	E	HS20	8.84	1.75	18.99	1	6	30
99	H	H20	14.63	1.47	26.91	1	7	0
100	H	H15	10.06	1.42	12.61	1	7	30

Bridge number	Bridge model information (inputs for ANN)				Moment rating factor (RF)			
	d_e (m)	t_s (mm)	f'_c (MPa)	F_y (MPa)	AASHTO (LRFD)	FEM	ANN (single)	ANN (committee)
1	0.20	165	20.69	344.75	0.930	1.112	1.122	1.137
2	−0.02	178	20.69	248.22	1.482	3.221	2.813	2.963
3	0.20	178	24.13	248.22	1.050	1.227	1.241	1.262
4	0.43	140	27.58	248.22	1.291	1.475	1.678	1.592
5	0.25	152	20.69	248.22	0.742	0.920	0.931	0.931
6	0.30	140	20.69	248.22	0.676	0.832	0.839	0.803
7	0.61	178	20.69	248.22	2.874	3.559	3.613	3.485
8	0.30	178	24.13	248.22	1.007	1.229	1.209	1.199
9	0.25	178	24.13	344.75	1.181	1.384	1.527	1.444
10	0.06	152	20.69	248.22	2.810	3.412	3.104	3.135
11	0.03	178	20.69	248.22	2.628	3.657	3.407	3.459
12	0.30	203	20.69	248.22	2.412	3.566	3.586	3.472
13	0.74	203	27.58	344.75	1.661	2.820	2.309	2.676
14	0.01	178	20.69	248.22	3.177	3.640	3.746	3.658
15	0.25	178	20.69	248.22	0.837	1.258	1.237	1.197
16	0.12	167	20.69	248.22	1.304	1.534	1.509	1.497
17	−0.11	178	27.58	248.22	2.563	3.178	3.108	3.105
18	0.08	203	27.58	344.75	2.137	2.720	2.693	2.654
19	0.28	127	20.69	248.22	1.075	1.795	1.828	1.791
20	0.25	165	27.58	344.75	1.689	2.040	1.975	1.902
21	0.25	203	27.58	248.22	1.401	2.682	2.676	2.702
22	0.07	127	20.69	248.22	1.232	1.425	1.425	1.405
23	0.23	203	27.58	344.75	2.953	3.134	2.955	3.037
24	0.86	152	27.58	248.22	1.327	1.779	1.762	1.864
25	0.28	165	27.58	344.75	1.734	2.211	2.177	2.144
26	0.28	216	24.13	248.22	2.255	2.421	2.674	2.490
27	0.06	178	20.69	344.75	1.880	2.277	2.209	2.163
28	0.15	178	17.24	248.22	1.717	2.055	2.055	2.128
29	0.16	178	20.69	248.22	2.379	2.775	2.714	2.741

(continued)

(continued)

Bridge number	Bridge model information (inputs for ANN)				Moment rating factor (RF)			
	d_e (m)	t_s (mm)	f'_c (MPa)	F_y (MPa)	AASHTO (LRFD)	FEM	ANN (single)	ANN (committee)
30	0.28	165	20.69	206.85	0.682	0.819	0.819	0.839
31	0.25	203	24.13	344.75	2.359	2.815	2.949	3.019
32	0.28	178	17.24	248.22	0.894	1.072	1.150	1.128
33	0.25	191	17.24	248.22	1.510	1.692	1.641	1.827
34	0.76	165	20.69	248.22	0.844	1.089	0.994	1.067
35	0.25	152	20.69	248.22	1.260	1.472	1.579	1.513
36	0.20	178	24.13	344.75	2.453	2.929	3.132	3.133
37	0.30	229	17.24	248.22	1.142	1.458	1.484	1.545
38	−0.20	178	27.58	248.22	1.643	1.912	2.084	1.989
39	0.23	178	17.24	248.22	0.675	0.826	0.802	0.840
40	0.28	203	20.69	344.75	1.032	1.594	1.687	1.770
41	0.20	165	17.24	248.22	0.903	1.069	1.036	1.100
42	0.30	178	20.69	248.22	1.327	1.599	1.627	1.644
43	0.04	203	20.69	248.22	2.214	2.666	2.495	2.635
44	0.03	203	27.58	344.75	2.247	3.081	2.947	2.978
45	0.18	178	17.24	248.22	0.986	1.192	1.234	1.238
46	0.69	191	24.13	248.22	1.836	2.405	2.391	2.507
47	0.30	140	17.24	248.22	0.655	0.782	0.776	0.828
48	0.10	178	27.58	344.75	2.126	2.521	2.547	2.486
49	0.20	203	20.69	248.22	2.523	3.620	3.615	3.605
50	0.00	178	27.58	248.22	2.225	2.452	2.556	2.557
51	0.03	178	27.58	248.22	2.145	2.704	2.730	2.746
52	0.23	140	27.58	248.22	1.317	1.449	1.432	1.468
53	0.36	178	27.58	344.75	1.304	1.716	1.764	1.771
54	0.66	165	27.58	248.22	0.808	1.339	1.331	1.270
55	0.76	203	20.69	344.75	0.747	1.565	1.354	1.464
56	0.05	152	20.69	248.22	1.443	1.647	1.627	1.606
57	0.25	165	17.24	227.54	1.113	1.345	1.411	1.429
58	0.23	178	27.58	248.22	1.309	1.601	1.551	1.553
59	0.25	165	20.69	248.22	1.709	1.898	1.781	1.795
60	0.23	152	17.24	206.85	1.590	1.741	1.779	1.780
61	0.28	165	27.58	248.22	1.619	1.876	1.926	1.933
62	0.28	165	20.69	248.22	1.716	1.804	1.774	1.761
63	0.28	165	20.69	248.22	1.385	1.561	1.543	1.567

(continued)

(continued)

Bridge number	Bridge model information (inputs for ANN)				Moment rating factor (RF)			
	d_e (m)	t_s (mm)	f'_c (MPa)	F_y (MPa)	AASHTO (LRFD)	FEM	ANN (single)	ANN (committee)
64	0.28	165	27.58	248.22	1.349	1.559	1.578	1.560
65	0.30	191	20.69	248.22	0.838	1.701	1.688	1.674
66	0.61	178	20.69	248.22	1.764	2.392	2.121	2.293
67	0.30	203	20.69	344.75	0.955	1.710	1.777	1.767
68	0.46	178	20.69	248.22	1.578	1.796	1.742	1.782
69	0.51	165	20.69	206.85	0.846	1.557	1.546	1.595
70	0.00	203	20.69	248.22	2.285	2.709	2.885	2.763
71	0.49	203	20.69	248.22	1.745	2.092	1.934	2.068
72	0.81	165	27.58	206.85	1.339	1.738	1.759	1.793
73	0.61	178	20.69	248.22	2.281	2.848	2.889	2.817
74	0.64	203	17.24	248.22	1.410	1.759	1.783	1.695
75	0.25	165	17.24	248.22	1.105	1.298	1.326	1.291
76	0.15	165	20.69	248.22	0.940	1.774	1.699	1.702
77	0.84	165	27.58	248.22	1.339	1.838	1.901	1.904
78	0.05	152	20.69	248.22	2.811	3.118	3.213	3.268
79	0.23	191	20.69	248.22	0.811	0.999	1.029	1.031
80	0.03	152	20.69	248.22	1.041	1.100	1.136	1.089
81	0.23	165	20.69	248.22	1.349	1.652	1.668	1.615
82	0.58	178	20.69	248.22	1.518	1.976	2.053	2.027
83	0.18	191	20.69	344.75	1.239	2.016	2.046	2.030
84	0.25	203	20.69	248.22	1.459	1.747	1.758	1.751
85	−0.11	178	17.24	248.22	1.661	2.019	1.965	1.965
86	0.79	178	17.24	206.85	1.323	1.573	1.644	1.661
87	0.81	178	24.13	248.22	1.102	1.407	1.450	1.420
88	0.41	152	20.69	248.22	1.435	1.647	1.528	1.610
89	0.23	127	17.24	206.85	1.135	2.144	2.241	2.212
90	0.74	178	20.69	248.22	0.933	1.130	1.016	1.117
91	0.30	165	20.69	227.54	1.308	1.523	1.423	1.408
92	0.46	165	27.58	248.22	1.133	2.399	2.275	2.337
93	0.53	191	24.13	248.22	2.949	3.162	2.784	2.643
94	0.28	165	17.24	206.85	0.699	0.952	1.051	1.078
95	0.23	178	20.69	248.22	1.222	1.447	1.409	1.370
96	0.66	178	27.58	248.22	1.332	2.351	2.471	2.477
97	0.25	178	20.69	248.22	1.518	1.739	1.790	1.784

(continued)

(continued)

Bridge number	Bridge model information (inputs for ANN)				Moment rating factor (RF)			
	d_e (m)	t_s (mm)	f'_c (MPa)	F_y (MPa)	AASHTO (LRFD)	FEM	ANN (single)	ANN (committee)
98	−0.11	178	20.69	248.22	2.358	2.947	3.054	3.043
99	0.22	165	20.69	206.85	0.857	1.039	1.066	1.053
100	0.30	191	20.69	206.85	1.148	1.650	1.649	1.640

References

1. AASHTO (2017) AASHTO LRFD bridge design specifications, 8th edn. Washington, DC
2. AASHTO (2018) Manual for bridge evaluation, 3rd edn.
3. Barker MG (2001) Quantifying field-test behavior for rating steel girder bridges. J Bridge Eng 6(4):254–261. http://doi.org/10.1061/(ASCE)1084-0702(2001)6:4(254)
4. Kim S, Nowak AS (1997) Load distribution and impact factors for I-girder bridges. J Bridg Eng 2(3):97–104
5. Nowak AS, Eom J (2001) Verification of girder distribution factors for steel girder bridges. MDOT 2000-0341 DIR. Michigan DOT, Lansing, MI
6. Sherman RJ, Hebdon MH, Lloyd JB (2020) Diagnostic load testing for improved accuracy of bridge load rating. J Perform Constr Facil 34(5):04020082. https://doi.org/10.1061/(ASCE)CF.1943-5509.0001483
7. Sofi FA, Steelman JS (2019) Nonlinear flexural distribution behavior and ultimate system capacity of skewed steel girder bridges. Eng Struct 197:109392. http://doi.org/10.1016/j.engstruct.2019.109392
8. Sofi FA, Steelman JS (2021) Using committees of artificial neural networks with finite element modeling for steel girder bridge load rating estimation. Structures 33:533–553. https://doi.org/10.1016/j.istruc.2021.04.056
9. Sofi FA (2017) Structural system-based evaluation of steel girder highway bridges and artificial neural network (ANN) implementation for bridge asset management. Doctoral dissertation, University of Nebraska-Lincoln, Lincoln, NE
10. Yousif Z, Hindi R (2007) AASHTO-LRFD live load distribution for beam-and-slab bridges: limitations and applicability. J Bridge Eng 12(6):765–773. http://doi.org/10.1061/(ASCE)1084-0702(2007)12:6(765)
11. Hearn G (2014) NCHRP synthesis 453: state bridge load posting processes and practices. Transportation Research Board, Washington, D.C.
12. Federal Highway Administration (FHWA), National Bridge Inventory (NBI) (2015) U.S. Department of Transportation, Washington, D.C. Retrieved December 31, 2015. www.fhwa.dot.gov/bridge/nbi.cfm
13. ANSYS Inc. (2019) ANSYS 19.0 [Computer Software]
14. Kathol S, Azizinamini A, Luedke J (1995) Strength capacity of steel girder bridges. Final Report RES1(0099) P469. Nebraska Department of Roads (NDOR), Lincoln, NE
15. Sofi FA, Steelman JS (2017) Parametric influence of bearing restraint on nonlinear flexural behavior and ultimate capacity of steel girder bridges. J Bridg Eng 22(7):04017033. https://doi.org/10.1061/(ASCE)BE.1943-5592.0001065
16. ANSYS Inc. (2019) ANSYS parametric design language guide, Release 2019, Canonsburg, PA

17. Sofi FA, Joo MR, Seetharaman R, Zakir M (2022) Compressive behavior and nonlinear load carrying capacity of multiple-shape concrete filled double-skin steel tubular columns. In: Maity D et al (eds) Recent advances in computational and experimental mechanics, vol I. Lecture notes in mechanical engineering. Springer, Singapore, pp 473–486
18. Zakir M, Sofi FA (2022) Experimental and nonlinear FE simulation-based compressive behavior of stiffened FRP-concrete-steel double-skin tubular columns with square outer and circular inner tubes. Eng Struct 260:114237. http://doi.org/10.1016/j.engstruct.2022.114237
19. Zakir M, Sofi FA, Behera S (2021) Nonlinear finite element analysis of circular stiffened FRP-concrete-steel double-skin tubular columns (DSTCs) and experimental compressive behavior of multiple DSTC shapes. Structures 34:3283–3299. https://doi.org/10.1016/j.istruc.2021.09.076
20. Zakir M, Sofi FA, Naqash JA (2021) Compressive testing and finite element analysis-based confined concrete model for stiffened square FRP-concrete-steel double-skin tubular columns. J Build Eng 44:103267. https://doi.org/10.1016/j.jobe.2021.103267
21. Joo MR, Sofi FA (2022) Forthcoming. Unified approach for estimating axial load capacity of concrete-filled double-skin steel tubular columns of multiple shapes using nonlinear FE models and artificial neural networks. Pract Period Struct Des Constr. https://doi.org/10.1061/(ASCE)SC.1943-5576.0000752
22. Willam KJ, Warnke EP (1974) Constitutive model for triaxial behaviour of concrete' concrete structures subjected to triaxial stresses. International Association for Bridges and Structural Engineering, Bergamo, Italy
23. Von Mises R (1913) Mechanik Der Festen Körper Im Plastisch-Deformablen Zustand. Mathematisch-Physikalische Klasse 582–592 (in German)
24. Kushvaha V, Anand Kumar S, Madhushri P, Sharma A (2020) Artificial neural network technique to predict dynamic fracture of particulate composite. J Compos Mater 54(22):3099–3108. https://doi.org/10.1177/0021998320911418
25. Sharma A, Mukhopadhyay T, Rangappa SM, Siengchin S, Kushvaha V (2022) Advances in computational intelligence of polymer composite materials: machine learning assisted modeling, analysis and design. Arch Comput Methods Eng. https://doi.org/10.1007/s11831-021-09700-9
26. Sharma A, Anand Kumar S, Kushvaha V (2020) Effect of aspect ratio on dynamic fracture toughness of particulate polymer composite using artificial neural network. Eng Fract Mech 228:106907. http://doi.org/10.1016/j.engfracmech.2020.106907
27. Sharma A, Kushvaha V (2020) Predictive modelling of fracture behaviour in silica-filled polymer composite subjected to impact with varying loading rates using artificial neural network. Eng Fract Mech 239:107328. https://doi.org/10.1016/j.engfracmech.2020.107328
28. Efron B, Tibshirani R (1986) Bootstrap methods for standard errors, confidence intervals, and other measures of statistical accuracy. Stat Sci 1(1):54–75
29. Principe JC, Euliano NR, Lefebvre WC (2000) Neural and adaptive systems: fundamentals through simulations. Wiley, New York
30. Marquardt DW (1963) An algorithm for least-squares estimation of nonlinear parameters. J Soc Ind Appl Math 11(2):431–441. https://doi.org/10.1137/0111030
31. MacKay DJC (1992) Bayesian interpolation. Neural Comput 4(3):415–447. https://doi.org/10.1162/neco.1992.4.3.415
32. Zokaie T (2000) AASHTO-LRFD live load distribution specifications. J Bridge Eng 5(2):131–138. http://doi.org/10.1061/(ASCE)1084-0702(2000)5:2(131)

Estimating Axial Load Capacity of Concrete-Filled Double-Skin Steel Tubular Columns of Multiple Shapes Using Genetic Algorithm-Optimized Artificial Neural Networks

Fayaz A. Sofi, Hazim Wani, Mohammad Zakir, and Manzoor A. Tantray

Notations

D_o	Outer diameter/edge dimension of outer steel tube;
D_i	Outer diameter/edge dimension of inner steel tube;
f_c'	Unconfined compressive strength of concrete (standard cylinder);
f_{yo}	Yield stress for outer steel tube;
f_{yi}	Yield stress for inner steel tube;
h	Column height;
I	Number of inputs;
m	Number of neurons or nodes per hidden layer;
O	Number of network outputs;
$P_{ult,ANN}$	ANN-based prediction for axial load capacity;
$P_{ult,exp}$	Experimental axial load capacity;
$P_{ult,FEM}$	FE-based axial load capacity;
$P_{ult,GA\text{-}ANN}$	GA-ANN model prediction for axial load capacity;
t_i	Thickness of inner steel tube;
t_o	Thickness of outer steel tube.

F. A. Sofi (✉) · H. Wani · M. Zakir · M. A. Tantray
Department of Civil Engineering, National Institute of Technology Srinagar, Hazratbal, Srinagar, Jammu and Kashmir 190006, India
e-mail: sofifayaz@nitsri.ac.in

M. Zakir
e-mail: zakirmohammad188_17@nitsri.net

M. A. Tantray
e-mail: matantray@nitsri.ac.in

V. Kushvaha et al. (eds.), *Machine Learning Applied to Composite Materials*, Composites Science and Technology, https://doi.org/10.1007/978-981-19-6278-3_9

1 Introduction

The new generation composite columns with concrete core sandwiched between inner and outer steel tubes are concrete-filled double-skin steel tubular (CFDST) columns. These composite columns are optimized for a high strength-to-weight ratio similar to an earlier generation of concrete-filled steel tubes (CFSTs). The CFDST columns offer many benefits compared to traditional reinforced concrete or steel columns, including better stiffness, improved energy absorption, residual axial load-carrying capacity and better ductility under dynamic and cyclic loads, provide higher strength compared to lesser weight, and have excellent fire resistance [1–7]. Both single and double-skin tubular columns have been widely used for civil engineering applications, such as bridge piers, columns and beams in high and medium-rise buildings, pile foundations, etc. [8]. The inner steel tube acts as the longitudinal reinforcement in these tubular columns, provides confinement to the concrete core and prevents its spalling [9–13]. For structural efficiency and aesthetic appeal, multiple shapes of CFDST columns have evolved using combinations of circular and square hollow steel sections, i.e., circle-circle (CC), circle-square (CS), square-square (SS) and square-circle (SC) cross-sections (Fig. 1). Accordingly, the CFDST columns of different cross-sectional shapes show varying behavior of confined concrete under axial compressive loading [8].

Each cross-sectional shape of CFDST columns has been extensively dealt with within the literature. In CC shape, Wei et al. [3] investigated the effects of variations in steel tube dimensions and sandwiched polymer concrete on CFDST column behavior. Experimental studies on CC shape [14, 15] showed confined concrete failure in CFDST columns due to local buckling deformations. Tao et al. [14] studied the failure modes of CFDST columns by varying the diameter-to-thickness ratio of steel tubes and the resulting column hollow ratio, i.e., inner-to-outer tube diameter ratio. For estimating the ultimate axial load capacity of CFDST columns, Tao et al. also proposed a simple equation based on the superposition of outer-tube-based CFST strength with inner steel tube strength. Other researchers studied the behavior of CC-shaped CFDST columns by performing detailed finite element (FE) simulations [2, 16–21] addressing the key issues for numerical simulations (such as

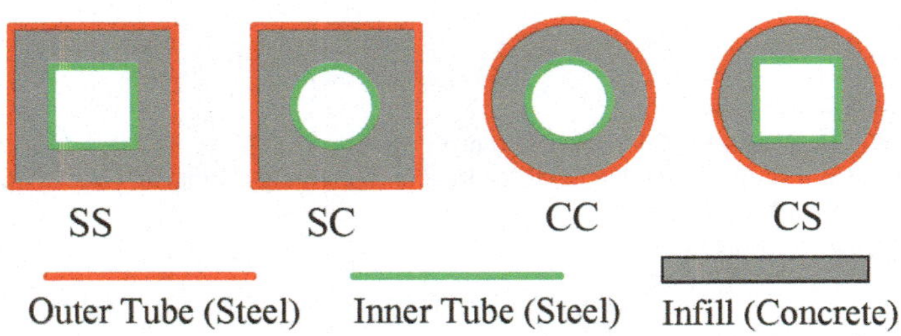

Fig. 1 CFDST column cross-sectional shapes (SS = square-square; SC = square-circle; CC = circle-circle; CS = circle-square)

constitutive modeling for confined concrete and steel, interactions between CFDST column components, etc.), and extended parametric variations in steel tube dimensions, concrete strengths, loading types, etc. Similarly, experimental studies on CS [22], SS [23] and SC [24] columns have investigated the effects of various parameters, such as hollow ratio, steel and concrete strengths, the width-to-thickness ratio of the inner tube, etc. on CFDST column behavior. The FE-based studies for CS [25], SC [18, 19, 26–28] and SS [1, 16, 29] cross-sectional shapes of CFDST columns have addressed various modeling issues, and extended influence of different geometry and material parameters ranges on the nonlinear column behavior.

Artificial neural networks (ANNs) are computational units with functional abstraction similar to biological neurons. ANNs are robust and adaptive in approximating complex, nonlinear multidimensional functions without the need for prior knowledge of nature or assumptions of the function. Figure 2 shows the multilayered feedforward architecture of ANNs, which is commonly used for mapping relationships between inputs and outputs [30–35]. Recently, ANNs have been used for predicting the structural responses of confined columns, including axial stress of FRP-confined concrete in circular columns [36], the ultimate axial load-carrying capacity of CFSTs [37–40] and CC-shaped CFDST columns [41]. Evolutionary algorithms, such as genetic algorithms (GA), can be efficiently used for the global solution search of the best network variables (i.e., weights and biases) in an ANN. The application of GAs requires randomly created individual solutions that are evaluated by a fitness function. The best pool of parents can then be used to generate the next population of solutions by recombination, crossover, and mutation inspired by biological genetics and survival of the fittest. These successive population generations can potentially lead to an optimal global solution for the optimization function evaluated by fitness criteria.

The main objective of the present study is to minimize the absolute maximum errors in estimating the ultimate axial load capacity of CC, CS, SC and SS-shaped CFDST columns using a hybrid GA and the multi-layered feedforward ANN model. A dataset of 171-CFDST columns (i.e., 51 of CC, 43 of CS, 38 of SC and 39 of SS shapes) was generated, i.e., 151 hypothetically generated and 20 existing columns using influential geometric and material properties referred to herein as governing parameters in this study. The hypothetical columns were randomly created using random combinations of governing parameter variations within their effective ranges. A detailed nonlinear FE methodology was first validated against documents test results of 40 CFDST columns specimens for different column cross-sections, such as CC [14, 15], CS [22], SC [24] and SS [23]. Then the validated FE modeling was used to analyze all 151-hypothetical columns to evaluate their ultimate axial load-bearing capacity. The proposed GA-ANN-based hybrid prediction model can be a more effective and practical tool for estimating the axial load-bearing capacity of CFDST columns of multiple cross-sectional shapes.

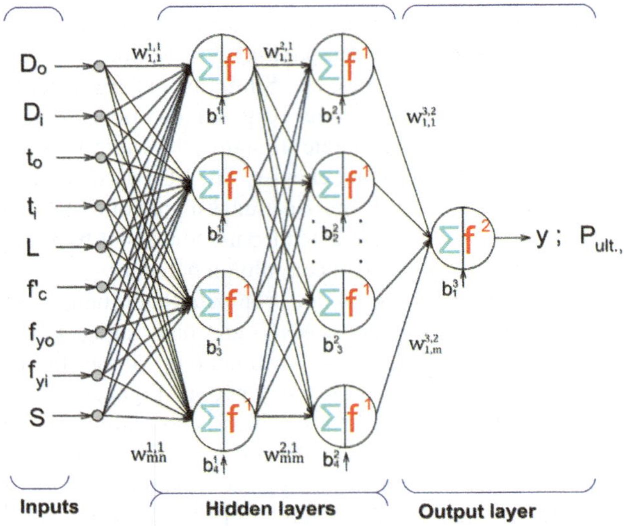

Fig. 2 Artificial neural network architecture (feedforward network): b = biases; $P_{ult.}$ = desired target; W = weights; y = output

2 FE Modeling and Simulation

The ultimate (peak) axial load capacity of CFDST columns used in ANN training was obtained from detailed nonlinear FE modeling for each cross-sectional shape using ANSYS 19.0 [42]. The schematic of the FE modeling framework is provided elsewhere [43]. In this consistent FE modeling procedure, custom-developed ANSYS parametric design language (APDL) [44]-based macros were used for automation in geometry creation, mapped meshing, the post-processing of element results, similar to Refs. [45, 46]. The constitutive material modeling for confined concrete and steel tubes was taken from documented literature on each cross-sectional column shape (details provided in Sect. 2.2 Constitutive Material Models). A summary of the adopted nonlinear FE simulation is provided in the following subsections.

2.1 Element Selections

Figure 3a shows the elements used to simulate different components of the CFDST column. The four-node shell elements (Shell 181), having 6 degrees of freedom (DOFs) per node, were used to idealize steel components (i.e., inner and outer tubes and top and bottom base plates). The concrete core was simulated by eight-node solid elements (Solid 185), having 3-DOFs per node. Interaction between steel tubes and concrete core was idealized with zero-length two-node rigid beams (MPC184)

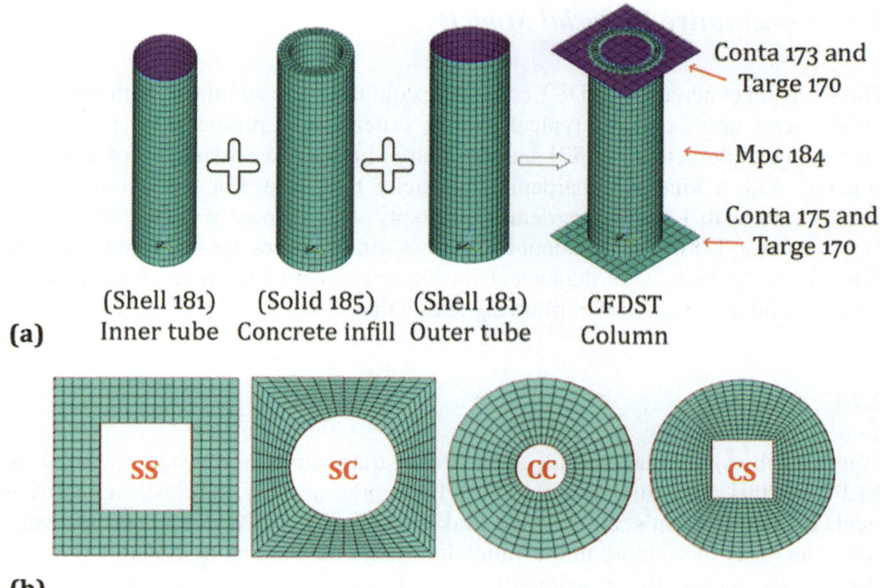

Fig. 3 FE modeling of CFDST columns: **a** element types; **b** mapped meshing in cross-sections

through kinematic relationships between DOFs of solid and shell elements. Surface-to-surface bonded type contact represented the interaction between steel endplates and concrete core at column end. This bonded type contact was simulated using 4-node contact elements (Conta 173), which were overlaid on the concrete cross-sections. The steel endplates provided the target surface using target elements (Targe 170). Similarly, node-to-surface type bonded contact simulated interaction between steel tubes and the endplates, using single-node contact elements (Conta 175) against target elements overlaid on steel endplates.

A mesh sensitivity analysis was performed for nonlinear FE analysis (NLFEA) results. The calibration study on the numerical models of CFDST columns of multiple shapes showed a maximum element size up to 15 mm in cross-sectional geometry, and 20 mm along the longitudinal axis of the column was sufficient for obtaining convergent results against documented experiments from literature (details provided in Sect. 2.4 FE Modeling Validation). The steel endplates were idealized with linear-elastic behavior and were meshed with a coarse element size of 15–20 mm. Figure 3b shows the details of mapped meshing used in multiple cross-sectional shapes of CFDST columns, i.e., CC, CS, SS and SC shapes.

2.2 Constitutive Material Models

The steel and concrete in CFDST composite columns were simulated with nonlinear axial stress–strain curves, typical failure criteria and plastic flow rules. The microplane failure theory [47] based failure surface was assumed for confined concrete with a kinematic-hardening plasticity flow rule. For steel components, nonlinearity with isotropic-hardening plasticity was adopted with the Von Mises yield criterion [48]. The simulation of stress–strain curves for steel and confined concrete in NLFEA were dependent on the cross-sectional shape of the CFDST columns and are given in the following subsections:

2.2.1 Steel

Figure 4a shows the elastoplastic axial stress–strain curve for the steel assumed for NLFEA in this study, similar to Ref. [49]. In the present study, the elastic modulus of steel (E_s) and Poisson's ratio (ν_s) were taken equal to 200 GPa and 0.3, respectively. After the onset of yielding in the validation cases, the hardening modulus (i.e., the slope of the second linear segment) of the stress–strain curve was taken equal to 1/100th of the initial elastic modulus (i.e., $E_1 = 0.01E_s$). In all other FE models (i.e., excluding validations cases), a perfectly elastoplastic stress–strain relationship (i.e., $E_1 = 0$ in Fig. 4a) was assumed in the NLFEA.

2.2.2 Concrete

CC and CS Shapes

The confined concrete axial stress–strain curve for CC and CS shapes of CFDST columns was simulated as given in Eqs. (1)–(9), similar to Ref. [21]. Figure 4b compares the axial stress–strain curves for unconfined and confined concrete idealized for CC and CS shapes of CFDST columns in this study. The confined concrete axial stress (f')–strain (ε) relationship, as proposed by Saenz [50], is given in Eqs. (1)–(8).

$$f' = \frac{E_{cc}\varepsilon}{1 + (R + R_E - 2)\left(\frac{\varepsilon}{\varepsilon_{cc}}\right) - (2R - 1)\left(\frac{\varepsilon}{\varepsilon_{cc}}\right)^2 + R\left(\frac{\varepsilon}{\varepsilon_{cc}}\right)^3} \text{ (MPa)} \quad (1)$$

where the confined concrete modulus, $E_{cc} = 4700\sqrt{f'_{cc}}$ (MPa), based on ACI-318 [51]. The parameters R_σ and R_ε are both taken equal to 4.0, as suggested by Hu and Schnobrich [52], and other variables R_E and R are given by Eqs. (2) and (3), respectively.

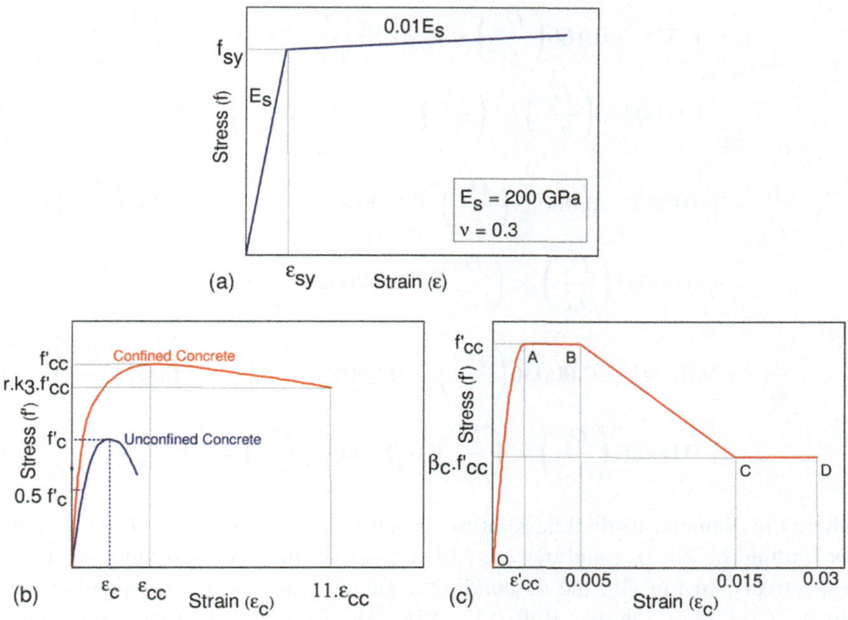

Fig. 4 Constitutive stress–strain models for CFDST columns: **a** steel tubes; **b** confined concrete in CC and CS shapes; **c** confined concrete in SS and SC shapes

$$R_E = \frac{E_{cc}\varepsilon_{cc}}{f'_{cc}} \tag{2}$$

$$R = \frac{R_E(R_\sigma - 1)}{(R_\varepsilon - 1)^2} - \frac{1}{R_\varepsilon} \tag{3}$$

Mander et al. [53] initially provided peak axial compressive stress (f'_{cc}) at the corresponding strain (ε_{cc}) in the confined concrete, as given by Eqs. (4) and (5):

$$f'_{cc} = f'_c + k_1 f_l \quad \text{(MPa)} \tag{4}$$

$$\varepsilon_{cc} = \varepsilon_{co}\left(1 + k_2\frac{f_l}{f'_c}\right) \tag{5}$$

where k_1 and k_2 are constants equal to 4.1 and 20.5, respectively, were recommended by Richart et al. [54]. The $f'_c = 0.8f_c$; unconfined cylinder strength of concrete and corresponding unconfined strain (ε_{co}) is taken equal to 0.003 in this study. Hu and Su [17] suggested effect of lateral confining pressure (f_l) as the minimum value obtained from Eqs. (6)–(8):

$$f_l = 8.525 - 0.166\left(\frac{D_o}{t_o}\right) - 0.00897\left(\frac{D_i}{t_i}\right) + 0.00125\left(\frac{D_o}{t_o}\right)^2 + 0.00246\left(\frac{D_o}{t_o}\right) \times \left(\frac{D_i}{t_i}\right) - 0.00550\left(\frac{D_i}{t_i}\right)^2 \geq 0 \quad (6)$$

$$\frac{f_l}{f_{yi}} = 0.01844 - 0.00055\left(\frac{D_o}{t_o}\right) + 0.00040\left(\frac{D_i}{t_i}\right) + 0.00001\left(\frac{D_o}{t_o}\right)^2 + 0.00001\left(\frac{D_o}{t_o}\right) \times \left(\frac{D_i}{t_i}\right) - 0.00002\left(\frac{D_i}{t_i}\right)^2 \quad (7)$$

$$\frac{f_l}{f_{yo}} = 0.01791 - 0.00036\left(\frac{D_o}{t_o}\right) - 0.00013\left(\frac{D_i}{t_i}\right) + 0.00001\left(\frac{D_o}{t_o}\right)^2 + 0.00001\left(\frac{D_o}{t_o}\right) \times \left(\frac{D_i}{t_i}\right) - 0.00002\left(\frac{D_i}{t_i}\right)^2 \geq 0 \quad (8)$$

where the diameter-to-thickness ratios of outer (D_o/t_o) and inner (D_i/t_i) are used for finding f_l. The f_{yo} and f_{yi} = yield stresses of the outer and inner steel tubes, respectively. In Fig. 4b, the descending segment's last point corresponds to axial stress, $f' = rk_3 f'_{cc}$ as per Refs. [17, 55]. The parameter, k_3 represents concrete degradation and is given by Eq. (9), Hu and Su [17]:

$$k_3 = 1.73916 - 0.00862\left(\frac{D_o}{t_o}\right) - 0.04731\left(\frac{D_i}{t_i}\right) - 0.00036\left(\frac{D_o}{t_o}\right)^2 + 0.00134\left(\frac{D_o}{t_o}\right)\left(\frac{D_i}{t_i}\right) - 0.00058\left(\frac{D_i}{t_i}\right)^2 \geq 0 \quad (9)$$

The reduction factor (r) was used as 1.0 for $f_c \leq 30$ MPa, [56], and 0.5 for $f_c \geq 100$ MPa, [57]. The linearly interpolated values were used for intermediate cube strengths in this study. For CS shaped CFDST columns, outer width (B_i) of the square inner tube was used instead of inner diameter (D_i) in Eqs. (1)–(9), similar to Hassanein et al. [25].

SS Shape

For NLFEA of SS-shaped DSTC columns, the confined concrete stress–strain model initially proposed by Mander et al. [53] and later modified by Zhao et al. [4] was used in this study. This analysis-oriented axial stress (f')–strain (ε) model for confined concrete in SS-shaped CFDST columns is given by Eqs. (10)–(16):

$$f' = \frac{f'_{cc}\left(\frac{\varepsilon}{\varepsilon_{cc}}\right)r_m}{r_m - 1 + \left(\frac{\varepsilon}{\varepsilon_{cc}}\right)^{r_m}} \quad (10)$$

$$f'_{cc} = f'_c\left[1 + \left(\frac{t_o}{B_0}\right)\left(\frac{f_{y0}}{f'_c}\right)\right] \quad (11)$$

$$\varepsilon_{cc} = \varepsilon_{co}\left[1 + 5\left(\frac{f'_{cc}}{f'_c} - 1\right)\right] \quad (12)$$

$$\varepsilon_{co} = 0.002 + 0.001\left(\frac{f'_c - 20}{80}\right) \quad (13)$$

$$r_m = \frac{\mathrm{E_c}}{\mathrm{E_c} - \mathrm{E_{sec}}} \quad (14)$$

$$\mathrm{E_c} = 3320\sqrt{f'_c} + 6900 \quad \text{(ACI 1984)} \quad (15)$$

$$\mathrm{E_{sec}} = \frac{f'_{cc}}{\varepsilon_{cc}} \quad (16)$$

SC Shape

Figure 4c shows this study's confined concrete axial stress–strain curve used in NLFEA of SC-shaped CFDST columns. This stress–strain model can be divided into four segments, i.e., OA, AB, BC, and CD (Fig. 4c), thus representing a modified version derived from the previous constitutive confined concrete model of the SS-shape (i.e., Eqs. 10–16). Segment OA can be obtained from the previous axial stress–strain model of the SS-shape for $f'_{cc} = \gamma_c f'_c$ [53]. Liang [58] suggested reduction factor (γ_c) for accounting effects of column size, concrete quality and the loading rate on confined concrete strength, as per Eq. (17):

$$\gamma_c = 1.85 D_c^{-0.135} \quad (0.85 \le \gamma_c \le 1.0) \quad (17)$$

where, D_c = depth of the concrete core in the column. Liang [58] also recommended strain (ε_{cc}) at f'_{cc} be used as per Eq. (18).

$$\varepsilon_{cc} = \begin{cases} 0.002 & \text{for } f'_{cc} \le 28 \text{ (MPa)} \\ 0.002 + \frac{f'_{cc} - 28}{54{,}000} & \text{for } 28 < f'_{cc} \le 82 \text{ (MPa)} \\ 0.003 & \text{for } f'_{cc} > 82 \text{ (MPa)} \end{cases} \quad (18)$$

The remaining segments of the axial stress–strain model (Fig. 4c), i.e., AB, BC and CD used in this study, were as also suggested by Liang [58] and given in Eq. (19):

$$f' = \begin{cases} f'_{cc} & \text{for } \varepsilon_{cc} < \varepsilon \le 0.005 \\ \beta_c f'_{cc} + 100(0.015 - \varepsilon)\left(f'_{cc} - \beta_c f'_{cc}\right) & \text{for } 0.005 < \varepsilon \le 0.015 \\ \beta_c f'_{cc} & \text{for } \varepsilon > 0.015 \end{cases} \quad (19)$$

where factor (β_c) accounts for the concrete ductility and is given by Eq. (20) as suggested by Liang [58]. The ultimate strain used in this study for confined concrete was limited to 0.03 in FE simulations of SC-shaped CFDST columns.

$$\beta_c = \begin{cases} 1.0 & \text{for } \frac{B_s}{t_o} \le 24 \\ 1.5 - \frac{1}{48}\frac{B_s}{t_o} & \text{for } 24 < \frac{B_s}{t_o} \le 48 \\ 0.5 & \text{for } \frac{B_s}{t_o} > 48 \end{cases} \quad (20)$$

2.3 Loading and Boundary Conditions

In this study, displacement-based loading and boundary conditions were applied to the endplate nodes of the numerical models. In the transverse directions (i.e., x and y global coordinate directions), both top and bottom endplates nodes were restrained against lateral translations. Whereas along the longitudinal axis of the columns (i.e., global z-direction), the translation was constrained at a single node located near the centroid of the bottom endplate. The compressive loading was applied in the longitudinal direction ($-z$ coordinate axis) by prescribing the displacement equal to 1/6th of the column height on the nodes of the top endplate in multiple load steps.

2.4 FE Modeling Validation

The nonlinear FE modeling methodology used in this study was first validated against the documented test results of 40 CFDST columns of multiple shapes, i.e., 16 specimens of CC [14, 15] and 8 specimens each of CS [22], SC [24] and SS [23] shapes. Figure 5a–d compare the typical experimental and analytical axial load–deflection characteristics of nominally identical specimens for SS, SC, CC, and CS shapes of CFDST columns, respectively. Table 1 summarizes the plotted results for the ultimate axial load capacity of these typical specimens of CFDST columns. The analytical and experimental axial load–deflection curves agree closely (Fig. 5). The mean ratio of FEM-to-experiment ultimate capacity on all 40 CFDST column specimens was close to 1.0 (Appendix 2). Thus, the validation of FE modeling provided the confidence that the adopted methodology in the present study has captured well the compressive behavior of multiple shapes of CFDST columns and can be reasonably extended for further generation of datasets for ANN model training and testing.

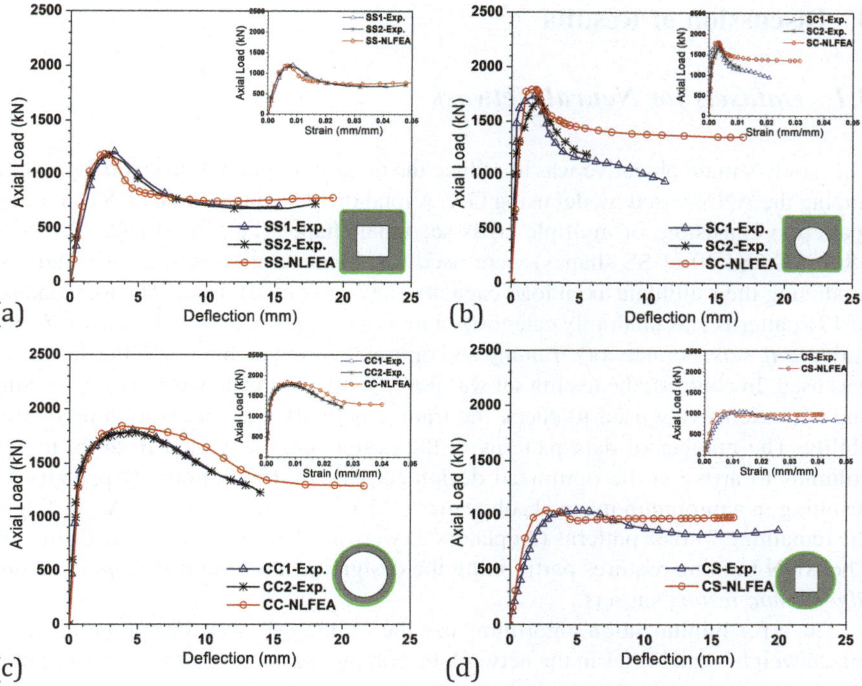

Fig. 5 FE modeling validations against documented axial load–deflection characteristics: **a** SS; **b** SC; **c** CC; **d** CS shapes

Table 1 Comparison of experimental and FE-based axial load capacity of typical CFDST columns

Column label	Ultimate axial-load capacity		$\frac{P_{ult,NLFEA}}{P_{ult.,Exp}}$
	$P_{ult.,\,Exp.}$ (kN)	$P_{ult.,\,NLFEA}$ (kN)	
CC1[a]	1790	1818	1.02
CC2[a]	1791		1.02
CS	1050	1074	1.02
SC1[a]	1725	1789	1.04
SC2[a]	1710		1.05
SS1[a]	1194	1189	1.00
SS2[a]	1210		0.98

[a] Nominally identical CFDST column specimens

3 Discussion of Results

3.1 *Datasets for Neural Network*

This study's main objective was to reduce the maximum prediction errors by reoptimizing the ANN-based model using GA. A total of 171-composite CFDST columns (patterns) consisting of multiple cross-sectional shapes (i.e., 51 of CC, 43 of CS, 38 of SC and 39 of SS shapes) were used for demonstration of the case study by predicting their ultimate axial load capacity (i.e., peak loads, $P_{ult.}$). The total dataset of 171 patterns was randomly categorized into two representative subsets, *the design* and *testing sets*. For network training and optimization of architecture, the design set was used. In contrast, the testing set was never shown to the network during training and was exclusively used to check the trained network's generalization prediction ability. The number of data patterns in the design set was varied from 20 to 120 columns to arrive at the optimized design set size of 100 patterns (Appendix 1), resulting in a minimum mean absolute error (MAE) on the testing set. Accordingly, the remaining 71 data patterns (Appendix 2) were used exclusively in the testing set. The ANN training requires partitioning the design set into *training*, *validation*, and *algorithmic-testing* subsets.

The error minimization algorithms use the training dataset for computing optimized weights and biases in the network by computing the gradients. The validation patterns were used for checking the network overfitting to training data during this optimization process. The network generalization performance is compared during training using algorithmic-testing set patterns. Clear evidence of the poor division of the design set into training, validation, and algorithmic-testing subsets can be seen when different iterations are needed for validation and algorithmic-testing data errors to reach a minimum [35]. The partitioning of the design set used proportions of 70% for training, 15% for validation and 15% for algorithmic-testing subsets in this study.

3.2 *ANN Model Predictions*

Optimizing the network architecture results in an efficient prediction model with enough network parameters, i.e., weights and biases, which can model the input–output mapping complexity of the problem in question. A too-large network size can sacrifice the network generalization ability by making the input–output mapping too specialized for the design set. A network with too few parameters results in poor input–output mapping ability.

In this study, two network architectures [*9-(m)-1* and *9-(m-m)-1*] consisting of one or two hidden layer(s) between an input layer (with 9-inputs) and an output layer with single output were optimized for hidden layer neurons (m). The network architectures were trained with selected error minimization algorithms (i.e., Levenberg–Marquardt LM, [59]; and Bayesian-regularization BR, [60]) to choose the best network based on mean square error (MSE) on the testing set. Out of the four trained network architectures, i.e., *9-(m)-1-BR, 9-(m-m)-1-BR, 9-(m)-1-LM, and 9-(m-m)-1-LM*, respectively, with *m* optimized to equal to 9, 4, 9 and 5 neurons in the hidden layer(s) were obtained. The optimized network, i.e., *9-(4–4)-1-BR*, showed about 7% mean absolute error for the testing set, which was the minimum among all considered network architectures. The best performing network *9-(4–4)-1-BR* was selected as the ANN prediction model in this study.

Figure 6 shows the performance of the ANN-based prediction model (i.e., *9-(4–4)-1-BR*) for estimating the ultimate axial load capacity ($P_{\mathrm{ult.,ANN}}$) of the CFDST columns of multiple cross-sectional shapes. The equal and fittings lines between the predicted and the actual (i.e., FE-based and experimental) ultimate axial load capacity are also shown in Fig. 6. The ANN-based prediction model showed a reasonably good prediction performance on CFDST columns of multiple shapes (i.e., CC, CS, SC and SS cross-sections). The absolute mean and maximum errors on the design set can be seen at about 3 and 18% for the design set, and 7 and 26% for the testing set, respectively (Fig. 6a). Between predicted and actual axial load capacity, the high R-square values of 0.9957 and 0.9587, respectively, for the design and testing sets demonstrate a close agreement between ANN model predictions and the actual CFDST column load-carrying capacities.

3.3 Hybrid GA-ANN Model Predictions

The genetic algorithm (GA) was used to reoptimize the network parameters (i.e., weights and biases) of the mapping function represented by the best-performing ANN model [*9-(4-4)-1-BR*]. This re-optimization process was based on the design set patterns to reduce the maximum prediction error in the testing set patterns. Accordingly, all network weights and biases in the mapping function represented by the ANN model were taken as parameters for optimization by GA. The fitness function was taken as the mean square error (*mse*) function on the design set. An initial population of 100 members represented by the weights and biases of the ANN model was created for GA by adding random values in the ranges of −0.03 to 0.03 to the original network parameters. The population members were ranked based on their fitness function values to determine the best-performing parents for generating a future population. The entire population of weights and biases were divided into elite and non-elite groups by having the top 30% ranked members in the former and the remaining members in the latter. The non-elite group was subjected to mathematical mutation and single-point crossover processes to generate the best-performing offspring on the fitness function evaluation. These processes regenerated the next

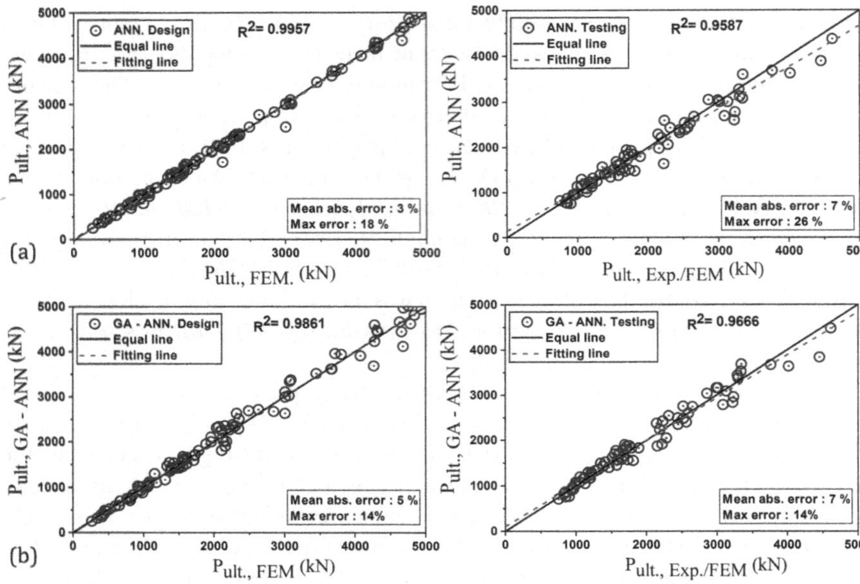

Fig. 6 Actual axial load capacity versus network model predictions for CFDST columns: **a** ANN model; **b** GA-ANN model; design set = 100 columns; testing set = 71 columns

generation population consisting of elite members from the previous generation and offsprings of the non-elite group. The whole procedure was repeated for about 10,000 iterations till a top-ranked member of weights and biases was obtained based on fitness evaluation (i.e., lowest *mse* value) on the design set.

Figure 6b shows the performance of the hybrid GA-ANN prediction model for estimating the ultimate axial load capacity ($P_{\text{ult., GA-ANN}}$) for CFDST columns of CC, CS, SC and SS cross-sectional shapes. The maximum errors for estimating $P_{\text{ult., GA-ANN}}$ in the GA-ANN prediction model have decreased from 26 to 14% for the testing set and 18 to 14% for the design set. For both ANN and GA-ANN prediction models, the mean absolute error for the testing set can be seen as similar at about 7% (Fig. 6b). Also, the R-square value has improved for the GA-ANN prediction model compared to the ANN model, i.e., from 0.9587 to 0.9666 for the testing set.

4 Conclusions

This study presented a hybrid genetic algorithm optimized-artificial neural network (GA-ANN)-based approach for estimating the ultimate axial load capacity of concrete-filled double-skin steel tubular (CFDST) columns for different combinations of circular and steel tube cross-sections, i.e., CC, CS, SC and SS shapes. A total dataset of 171-CFDST columns (i.e., 51 of CC, 43 of CS, 38 of SC and 39 of SS shapes) was analyzed using nonlinear finite element analysis to obtain their ultimate axial load capacity and was used in the case study for demonstration. The governing inputs in both ANN and GA-ANN prediction models were geometry and material properties of CFDST columns, i.e., inner and outer steel tube diameter, thickness, cross-sectional shape, height and yield strength, and unconfined compressive strength of sandwiched concrete. Based on the results of this study, it is concluded that a hybrid GA-ANN prediction model is more accurate and can significantly reduce the maximum prediction errors than the ANN model in estimating the axial load capacity of CFDST columns of multiple cross-sectional shapes (i.e., about 26 to 14% error reduction in this study). The proposed GA-ANN prediction model can be a potentially practical tool for estimating the axial load capacity of CFDST columns.

Appendix 1

The design set of 100 columns used for ANN design (training, validation and algorithmic testing)

Column number	Inputs for ANN									$P_{ult.,FEM}$ (kN)	$P_{ult.,ANN}$ (kN)	$P_{ult.,GA\text{-}ANN}$ (kN)
	Cross-sectional shape[a]	D_o (mm)	t_o (mm)	D_i (mm)	t_i (mm)	L (mm)	f_c' (MPa)	f_{yo} (MPa)	f_{yi} (MPa)			
1	SS	195	3.5	140	3.5	700	20	400	500	1558	1640	1687
2	CC	125	2.5	70	3.5	500	40	300	350	956	932	934
3	CS	100	2.5	40	1.5	300	24	350	400	520	532	545
4	SC	230	3.5	165	6.5	700	60	300	500	4075	4058	3918
5	CS	125	2.5	50	2.5	400	64	350	400	1181	1167	1116
6	SS	60	1.5	40	1.5	300	48	300	300	265	264	269
7	CS	140	3.5	75	2.5	400	32	250	400	928	906	961
8	SS	90	1.5	40	1.5	400	32	350	300	467	469	495
9	CS	175	2.5	60	1.5	700	20	450	350	1076	1096	1128
10	SS	190	2.5	125	3.5	700	36	300	350	1157	1153	1307
11	SC	260	6.5	105	3.5	1000	44	500	250	4940	4963	4948
12	CS	270	4.5	90	3.5	1400	60	300	300	4785	4792	4622
13	SS	150	2.5	90	2.5	500	40	250	350	986	806	895
14	SC	120	1.5	65	2.5	300	52	350	500	1052	1035	1017
15	SC	140	3.5	75	1.5	300	48	500	400	1572	1628	1645
16	SS	125	2.5	30	1.5	300	36	500	300	1069	1037	1082
17	SS	205	4.5	115	4.5	800	28	250	400	1726	1749	1708

(continued)

(continued)

Column number	Inputs for ANN									$P_{ult.,FEM}$ (kN)	$P_{ult.,ANN}$ (kN)	$P_{ult.,GA\text{-}ANN}$ (kN)
	Cross-sectional shape[a]	D_o (mm)	t_o (mm)	D_i (mm)	t_i (mm)	L (mm)	f_c' (MPa)	f_{yo} (MPa)	f_{yi} (MPa)			
18	SS	85	1.5	45	1.5	300	24	450	450	448	490	509
19	CC	305	5.5	180	4.5	700	52	400	300	4683	4671	4975
20	CC	75	1.5	30	1.5	300	60	300	450	410	392	370
21	SC	190	2.5	90	4.5	1200	60	450	250	2845	2833	2680
22	CS	325	6.5	160	3.5	1400	48	300	300	4679	4679	4127
23	SS	260	6.5	125	3.5	800	20	400	300	3073	3097	3343
24	CS	160	3.5	55	1.5	400	48	350	350	1504	1422	1440
25	SC	210	3.5	60	1.5	900	52	350	400	3004	2999	2998
26	CC	270	4.5	140	5.5	500	40	300	250	3711	3722	3962
27	SC	85	1.5	50	2.5	300	32	300	350	438	491	423
28	CC	115	2.5	45	1.5	300	60	450	500	977	1022	1025
29	CC	100	2.5	65	2.5	400	48	450	300	668	703	714
30	CC	330	5.5	115	2.5	800	32	400	250	4277	4346	4596
31	CS	195	3.5	115	2.5	800	32	350	300	1364	1457	1549
32	SC	210	3.5	115	2.5	800	28	500	250	2220	2210	2280
33	SS	105	1.5	50	2.5	300	20	400	400	594	554	599

(continued)

(continued)

Column number	Inputs for ANN									$P_{ult.,FEM}$ (kN)	$P_{ult.,ANN}$ (kN)	$P_{ult.,GA\text{-}ANN}$ (kN)
	Cross-sectional shape[a]	D_o (mm)	t_o (mm)	D_i (mm)	t_i (mm)	L (mm)	f_c' (MPa)	f_{yo} (MPa)	f_{yi} (MPa)			
34	CC	245	3.5	160	4.5	900	20	300	300	1958	1955	2109
35	CC	85	1.5	40	1.5	300	32	350	500	399	412	412
36	SC	295	6.5	140	5.5	900	24	400	500	4631	4593	4669
37	SC	180	4.5	60	1.5	500	20	500	400	1730	1767	1730
38	SC	75	1.5	40	1.5	300	32	350	350	344	376	332
39	SC	175	3.5	65	2.5	600	56	500	300	2631	2774	2725
40	CC	280	3.5	140	3.5	1300	40	300	300	3450	3481	3519
41	SC	115	2.5	70	1.5	300	44	350	250	816	786	793
42	SS	295	4.5	175	3.5	1200	32	300	350	2320	2294	2635
43	SS	305	5.5	125	2.5	600	32	350	350	3088	3012	3385
44	SS	330	5.5	165	6.5	900	52	400	500	4267	4261	3692
45	SS	200	2.5	115	2.5	900	40	400	500	1432	1501	1548
46	CC	150	2.5	45	1.5	400	24	300	350	918	1010	1044
47	CS	150	2.5	40	1.5	500	52	450	350	1480	1484	1438

(continued)

(continued)

Column number	Inputs for ANN									$P_{ult.,FEM}$ (kN)	$P_{ult.,ANN}$ (kN)	$P_{ult.,GA\text{-}ANN}$ (kN)
	Cross-sectional shape[a]	D_o (mm)	t_o (mm)	D_i (mm)	t_i (mm)	L (mm)	f_c' (MPa)	f_{yo} (MPa)	f_{yi} (MPa)			
48	CC	275	5.5	135	4.5	1300	56	350	400	4839	4836	4818
49	SS	270	4.5	160	3.5	700	56	350	250	2069	2079	2343
50	CS	140	3.5	90	4.5	400	60	400	300	1399	1423	1426
51	CS	100	2.5	30	1.5	300	52	400	400	765	801	771
52	SC	220	5.5	130	6.5	700	36	400	500	3666	3723	3620
53	CC	150	2.5	105	3.5	400	24	400	350	1064	972	1050
54	CS	150	2.5	90	2.5	700	32	350	300	935	1003	1030
55	CS	220	5.5	140	3.5	600	44	250	350	1963	1979	2016
56	CC	200	2.5	140	5.5	800	52	350	300	2302	2349	2362
57	SC	260	6.5	125	2.5	800	48	300	300	4310	4259	4478
58	SS	205	4.5	165	6.5	400	56	450	300	2179	2179	2023
59	CC	180	4.5	125	2.5	700	28	450	400	1453	1369	1472
60	SC	175	2.5	130	6.5	500	20	250	400	1445	1457	1431
61	CC	105	1.5	50	2.5	600	40	250	450	654	657	621
62	CS	270	4.5	100	2.5	1300	32	450	300	3061	3030	3029

(continued)

(continued)

Column number	Inputs for ANN									$P_{ult.,FEM}$ (kN)	$P_{ult.,ANN}$ (kN)	$P_{ult.,GA\text{-}ANN}$ (kN)
	Cross-sectional shape[a]	D_o (mm)	t_o (mm)	D_i (mm)	t_i (mm)	L (mm)	f_c' (MPa)	f_{yo} (MPa)	f_{yi} (MPa)			
63	SC	100	2.5	30	1.5	300	52	350	300	808	842	775
64	SC	210	3.5	140	3.5	500	64	250	400	3004	2508	2642
65	SS	280	3.5	135	4.5	1100	64	300	350	2159	2162	1914
66	SC	245	3.5	165	6.5	800	32	300	300	3004	3015	3115
67	CS	295	6.5	100	2.5	700	28	450	500	3802	3771	3952
68	CC	260	6.5	115	4.5	800	48	350	500	4655	4393	4453
69	CS	115	1.5	50	2.5	300	28	400	450	808	706	723
70	CC	295	4.5	90	4.5	1000	28	350	250	4277	4213	4232
71	SS	245	3.5	135	4.5	500	40	500	250	2040	2094	2318
72	CC	140	3.5	45	1.5	500	60	450	350	1514	1495	1475
73	SS	265	3.5	115	2.5	700	48	400	450	2155	2161	2360
74	SS	230	3.5	135	4.5	400	32	500	300	2111	2064	2265
75	SS	150	2.5	105	3.5	600	32	500	450	1302	1245	1173
76	SC	200	2.5	140	3.5	600	60	400	300	2263	2283	2389
77	SS	140	2.5	40	1.5	500	56	350	400	1377	1394	1370

(continued)

(continued)

Column number	Inputs for ANN									$P_{ult.,FEM}$ (kN)	$P_{ult., ANN}$ (kN)	$P_{ult.,GA-ANN}$ (kN)
	Cross-sectional shape[a]	D_o (mm)	t_o (mm)	D_i (mm)	t_i (mm)	L (mm)	f_c' (MPa)	f_{yo} (MPa)	f_{yi} (MPa)			
78	SS	160	3.5	90	2.5	400	40	500	350	1529	1453	1555
79	CC	165	2.5	55	1.5	400	44	500	300	1564	1561	1592
80	CS	165	2.5	100	4.5	500	60	350	300	1606	1674	1672
81	CC	195	3.5	115	4.5	700	60	300	250	2280	2320	2341
82	SC	190	2.5	115	4.5	600	24	400	350	1781	1823	1817
83	CC	160	3.5	60	1.5	400	48	350	350	1384	1404	1447
84	SC	295	6.5	100	2.5	1100	24	500	350	4322	4339	4440
85	CS	115	2.5	45	1.5	300	56	350	350	848	851	848
86	CS	150	2.5	90	4.5	500	60	350	350	1554	1489	1469
87	SC	195	3.5	125	2.5	600	40	400	250	2122	2035	2099
88	CC	190	2.5	90	4.5	600	56	350	250	2350	2322	2296
89	CS	195	3.5	115	4.5	800	48	300	350	2139	2066	2021
90	SC	260	6.5	135	4.5	700	48	400	250	4759	4877	5018
91	CS	195	3.5	75	2.5	500	24	250	500	1333	1394	1477
92	CC	265	3.5	100	2.5	1000	52	300	300	3655	3621	3619

(continued)

(continued)

Column number	Inputs for ANN									$P_{ult.,FEM}$ (kN)	$P_{ult., ANN}$ (kN)	$P_{ult.,GA\text{-}ANN}$ (kN)
	Cross-sectional shape[a]	D_o (mm)	t_o (mm)	D_i (mm)	t_i (mm)	L (mm)	f_c' (MPa)	f_{yo} (MPa)	f_{yi} (MPa)			
93	SS	220	5.5	115	2.5	500	32	450	400	2492	2501	2695
94	CS	160	3.5	60	1.5	700	48	500	500	1578	1684	1615
95	CS	160	3.5	65	2.5	600	52	300	350	1606	1558	1526
96	SC	180	4.5	115	2.5	400	32	450	500	2108	1730	1820
97	SS	250	4.5	125	2.5	500	32	400	500	2354	2351	2512
98	SC	175	3.5	70	1.5	700	36	400	400	1883	1971	1961
99	CS	165	2.5	65	2.5	700	48	400	500	1706	1674	1599
100	SS	140	2.5	75	2.5	600	60	500	400	1387	1375	1409

[a] Column cross-sectional shape codes used as ANN inputs, CC = 1; CS = 2; SS = 3; SC = 4

Appendix 2

The testing set of 71 columns (51 FE-analysed and 20 actual tested columns) used for testing ANN and GA-ANN model performances

Column number (label)	Inputs for ANN									$P_{ult.,FEM/Exp}$ (kN)	$P_{ult., ANN}$ (kN)	$P_{ult., GA\text{-}ANN}$ (kN)
	Cross-sectional shape[a]	D_o (mm)	t_o (mm)	D_i (mm)	t_i (mm)	L (mm)	f_c' (MPa)	f_{yo} (MPa)	f_{yi} (MPa)			
1	SS	195	3.5	70	1.5	600	32	250	300	1675	1475	1672
2	SC	260	6.5	140	3.5	900	44	400	400	4612	4376	4480
3	CC	250	4.5	160	4.5	600	32	400	400	3223	2605	2836
4	SS	250	4.5	140	3.5	700	60	350	400	2154	1984	1876
5	CC	245	3.5	100	2.5	700	48	400	300	3125	3006	3088
6	CC	175	3.5	130	6.5	600	28	500	350	1890	1800	1835
7	CS	115	2.5	40	1.5	300	60	400	350	961	960	939
8	CS	120	1.5	45	1.5	300	56	300	250	851	768	767
9	CC	205	4.5	135	4.5	800	48	300	300	2146	2296	2378
10	CC	195	3.5	70	1.5	500	64	400	400	2447	2326	2356
11	CS	195	3.5	105	3.5	700	28	300	400	1736	1509	1580
12	CS	190	2.5	115	4.5	500	32	450	300	1516	1526	1626
13	SC	250	4.5	165	5.5	900	40	350	500	4018	3622	3639
14	CC	140	3.5	60	1.5	400	52	400	400	1205	1214	1245
15	CC	250	5.5	135	4.5	800	32	300	400	3013	3006	3164
16	SC	100	2.5	40	1.5	300	48	300	500	754	829	721
17	CS	295	4.5	90	4.5	1100	24	400	400	4452	3890	3841

(continued)

(continued)

Column number (label)	Inputs for ANN									$P_{ult.,FEM/Exp}$ (kN)	$P_{ult.,ANN}$ (kN)	$P_{ult.,GA\text{-}ANN}$ (kN)
	Cross-sectional shape[a]	D_o (mm)	t_o (mm)	D_i (mm)	t_i (mm)	L (mm)	f_c' (MPa)	f_{yo} (MPa)	f_{yi} (MPa)			
18	SS	210	3.5	115	2.5	600	32	400	300	2223	1642	1918
19	SS	175	2.5	55	1.5	700	32	450	400	1463	1350	1413
20	CS	220	5.5	105	3.5	500	60	300	400	3081	2693	2791
21	CS	140	3.5	75	1.5	500	32	500	500	994	990	1032
22	CC	175	3.5	75	1.5	500	40	350	350	1416	1464	1528
23	CS	230	3.5	70	3.5	800	20	400	500	2281	2070	2052
24	CS	250	5.5	105	3.5	500	44	300	300	3286	3269	3437
25	CS	180	4.5	60	1.5	900	52	300	300	1759	1937	1883
26	CC	100	2.5	50	2.5	300	56	500	350	849	878	876
27	CS	160	3.5	55	1.5	400	48	350	350	1379	1422	1440
28	CC	205	4.5	135	4.5	500	24	250	250	1562	1612	1776
29	CC	175	2.5	90	2.5	400	20	300	450	1127	1094	1187
30	CS	205	4.5	65	2.5	600	40	350	300	2187	2223	2244
31	SC	225	4.5	90	2.5	500	28	250	450	2314	2434	2546
32	SS	295	4.5	115	2.5	800	32	450	500	3230	2779	2962
33	SS	150	2.5	75	1.5	600	48	450	350	1269	1259	1335

(continued)

(continued)

Column number (label)	Inputs for ANN									$P_{ult.,FEM/Exp}$ (kN)	$P_{ult., ANN}$ (kN)	$P_{ult., GA-ANN}$ (kN)
	Cross-sectional shape[a]	D_o (mm)	t_o (mm)	D_i (mm)	t_i (mm)	L (mm)	f_c' (MPa)	f_{yo} (MPa)	f_{yi} (MPa)			
34	SS	295	6.5	105	3.5	600	40	400	300	3761	3683	3673
35	SS	115	2.5	30	1.5	300	40	450	350	949	956	980
36	SC	210	3.5	130	6.5	700	48	250	300	2857	3049	3034
37	SS	210	3.5	125	3.5	500	60	450	300	1811	1490	1563
38	CC	225	4.5	75	2.5	700	36	350	300	2599	2540	2604
39	CC	245	3.5	90	2.5	800	28	350	400	2527	2400	2491
40	CC	150	2.5	90	3.5	500	24	300	450	1127	1015	1063
41	CC	175	3.5	60	1.5	700	40	400	400	1548	1698	1704
42	SC	165	2.5	105	3.5	900	28	250	400	1349	1581	1503
43	CC	175	3.5	75	2.5	400	40	300	350	1588	1512	1578
44	SS	175	2.5	90	3.5	700	36	250	350	1198	1139	1177
45	CC	180	4.5	70	3.5	700	56	450	250	2562	2449	2408
46	CS	295	6.5	100	2.5	1200	24	400	400	3341	3597	3688
47	CC	220	5.5	115	2.5	900	52	350	350	2650	2680	2748

(continued)

(continued)

Column number (label)	Inputs for ANN									$P_{ult.,FEM/Exp}$ (kN)	$P_{ult., ANN}$ (kN)	$P_{ult., GA\text{-}ANN}$ (kN)
	Cross-sectional shape[a]	D_o (mm)	t_o (mm)	D_i (mm)	t_i (mm)	L (mm)	f_c' (MPa)	f_{yo} (MPa)	f_{yi} (MPa)			
48	CS	175	3.5	75	2.5	400	28	500	300	1576	1382	1459
49	SS	275	5.5	125	2.5	900	60	250	350	2234	2596	2429
50	CC	165	2.5	100	2.5	500	48	300	250	1173	1241	1300
51	SS	250	5.5	140	3.5	500	40	500	350	2990	3042	3161
52[b] (cc2a, b)	CC	180	3	48	3	540	37.92	275.9	396.1	1791	1850	1836
53[b] (cc3a, b)	CC	180	3	88	3	540	37.92	275.9	370.2	1649	1538	1589
54[b] (cc5a, b)	CC	114	3	58	3	342	37.92	294.5	374.5	901	757	778
55[b] (cc6a, b)	CC	240	3	114	3	720	37.92	275.9	294.5	2441	2385	2492
56[b] (cc7a, b)	CC	300	3	165	3	900	37.92	275.9	320.5	3299	3125	3383
57[b](C4C7)	CC	114.20	3	48.4	2.80	400	63.40	430	425	1110	1207	1178
58[b] (C5C8)	CC	165.10	3.50	101.8	3.10	400	63.40	433	410	1705	1744	1805
59[b] (C6C8)	CC	165.30	2.90	101.8	3.10	400	63.40	395	410	1605	1654	1711
60[c] (C4S1)	CS	114.41	2.96	40.14	1.80	401	63.47	430	492	1050	1164	1094
61[c] (C5S2)	CS	139.67	3.42	65.17	2.47	399	63.47	379	392	1315	1401	1373
62[c] (C6S2)	CS	139.67	2.94	65.17	2.47	399	63.47	357	392	1215	1327	1302

(continued)

(continued)

Column number (label)	Inputs for ANN									$P_{ult.,FEM/Exp}$ (kN)	$P_{ult.,ANN}$ (kN)	$P_{ult.,GA\text{-}ANN}$ (kN)
	Cross-sectional shape[a]	D_o (mm)	t_o (mm)	D_i (mm)	t_i (mm)	L (mm)	f_c' (MPa)	f_{yo} (MPa)	f_{yi} (MPa)			
63[c] (C7S3)	CS	165.18	3.44	75.29	2.90	400	63.47	433	400	1690	1949	1914
64[c] (C8S3)	CS	165.18	2.90	75.29	2.90	400	63.47	395	400	1665	1826	1797
65[d] (CS4S5A, B)	SS	100.46	2.06	50	2.44	375	58.70	453	477	830	788	834
66[e] (scc2-1, 2)	SC	120	3	32	3	360	37.40	275.9	422.3	1057	1299	1201
67[e] (scc3-1, 2)	SC	120	3	58	3	360	37.40	275.9	374.5	995	1142	1089
68[e] (scc4-1, 2)	SC	120	3	88	3	360	37.40	275.9	370.2	933	964	912
69[e] (scc5-1, 2)	SC	180	3	88	3	540	37.40	275.9	370.2	1718	1865	1866
70[e] (scc6-1, 2)	SC	240	3	114	3	720	37.40	275.9	294.5	2520	2547	2752
71[e] (scc7-1, 2)	SC	300	3	165	3	900	37.40	275.9	320.5	3335	3084	3528

[a] Column cross-sectional shape codes used as ANN inputs, CC = 1; CS = 2; SS = 3; SC = 4

[b] Tao et al. [14]

[c] Elchalakani et al. [22]

[d] Zhao and Grzebieta [23]

[e] Han et al. [24]

Columns 52 to 71 are actual tested specimens; $P_{ult.,Exp.}$ = experimentally observed peak axial load

References

1. Liang W, Dong J, Wang Q (2019) Mechanical behaviour of concrete-filled double-skin steel tube (CFDST) with stiffeners under axial and eccentric loading. Thin-Walled Struct 138:215–230. http://doi.org/10.1016/j.tws.2019.02.002
2. Wang F, Young B, Gardner L (2019) Compressive testing and numerical modelling of concrete-filled double skin CHS with austenitic stainless steel outer tubes. Thin-Walled Struct 141:345–359. http://doi.org/10.1016/j.tws.2019.04.003
3. Wei S, Mau ST, Vipulanandan C, Mantrala SK (1995) Performance of new sandwich tube under axial loading: experiment. J Struct Eng 121(12):1806–1814. https://doi.org/10.1061/(ASCE)0733-9445(1995)121:12(1806)
4. Zhao X, Han B, Grzebieta RH (2002a) Plastic mechanism analysis of concrete-filled double-skin (SHS inner and SHS outer) stub columns. Thin-Walled Struct 40(10):815–833. https://doi.org/10.1016/S0263-8231(02)00030-7
5. Zhao XL, Grzebieta RH, Ukur A, Elchalakani M (2002b) Tests of concrete-filled double skin (SHS outer and CHS inner) composite stub columns. Adv Steel Struct 1:567–574
6. Tao Z, Han L-H, Wang ZB (2005) Experimental behaviour of stiffened concrete-filled thin-walled hollow steel structural (HSS) stub columns. J Constr Steel Res 61(7):962–983. https://doi.org/10.1016/j.jcsr.2004.12.003
7. Joo MR, Sofi FA (2022) Forthcoming. Unified approach for estimating axial load capacity of concrete-filled double-skin steel tubular columns of multiple shapes using nonlinear FE models and artificial neural networks. Pract Period Struct Des Constr. https://doi.org/10.1061/(ASCE)SC.1943-5576.0000752
8. Sofi FA, Joo MR, Seetharaman R, Zakir M (2022) Compressive behavior and nonlinear load carrying capacity of multiple-shape concrete filled double-skin steel tubular columns. In: Recent advances in computational and experimental mechanics, vol I. Lecture notes in mechanical engineering. Springer, Singapore, pp 473–486
9. Sofi FA, Zakir M, Naqash JA (2022) Experimentally verified behavior of multiple-shape hybrid FRP-concrete-steel double-skin tubular columns under axial compression. In: Recent advances in computational and experimental mechanics, vol I. Lecture notes in mechanical engineering. Springer, Singapore, pp 167–175
10. Zakir M, Sofi FA (2022) Experimental and nonlinear FE simulation-based compressive behavior of stiffened FRP-concrete-steel double-skin tubular columns with square outer and circular inner tubes. Eng Struct 260:114237. http://doi.org/10.1016/j.engstruct.2022.114237
11. Zakir M, Sofi FA, Behera S (2021) Nonlinear finite element analysis of circular stiffened FRP-concrete-steel double-skin tubular columns (DSTCs) and experimental compressive behavior of multiple DSTC shapes. Structures 34:3283–3299. https://doi.org/10.1016/j.istruc.2021.09.076
12. Zakir M, Sofi FA, Naqash JA (2021) Experimentally verified behavior and confinement model for concrete in circular stiffened FRP-concrete-steel double-skin tubular columns. Structures 33:1144–1157. https://doi.org/10.1016/j.istruc.2021.05.010
13. Zakir M, Sofi FA, Naqash JA (2021) Compressive testing and finite element analysis-based confined concrete model for stiffened square FRP-concrete-steel double-skin tubular columns. J Build Eng 44:103267. https://doi.org/10.1016/j.jobe.2021.103267
14. Tao Z, Han L-H, Zhao X-L (2004) Behaviour of concrete-filled double skin (CHS inner and CHS outer) steel tubular stub columns and beam-columns. J Constr Steel Res 60(8):1129–1158. https://doi.org/10.1016/j.jcsr.2003.11.008
15. Zhao X-L, Grzebieta R, Elchalakani M (2002) Tests of concrete-filled double skin CHS composite stub columns. Steel Compos Struct 2(2):129–146. http://doi.org/10.12989/scs.2002.2.2.129
16. Han L-H, Li Y-J, Liao F-Y (2011) Concrete-filled double skin steel tubular (CFDST) columns subjected to long-term sustained loading. Thin-Walled Struct 49(12):1534–1543. http://doi.org/10.1016/j.tws.2011.08.001

17. Hu HT, Su FC (2011) Nonlinear analysis of short concrete-filled double skin tube columns subjected to axial compressive forces. Mar Struct 24(4):319–337. http://doi.org/10.1016/j.marstruc.2011.05.001
18. Huang H, Han L-H, Tao Z, Zhao X-L (2010) Analytical behaviour of concrete-filled double skin steel tubular (CFDST) stub columns. J Constr Steel Res 66(4):542–555. http://doi.org/10.1016/j.jcsr.2009.09.014
19. Huang H, Han L-H, Zhao X-L (2013) Investigation on concrete filled double skin steel tubes (CFDSTs) under pure torsion. J Constr Steel Res 90:221–234. https://doi.org/10.1016/j.jcsr.2013.07.035
20. Liang QQ (2017) Nonlinear analysis of circular double-skin concrete-filled steel tubular columns under axial compression. Eng Struct 131:639–650. http://doi.org/10.1016/j.engstruct.2016.10.019
21. Pagoulatou M, Sheehan T, Dai XH, Lam D (2014) Finite element analysis on the capacity of circular concrete-filled double-skin steel tubular (CFDST) stub columns. Eng Struct 72:102–112. http://doi.org/10.1016/j.engstruct.2014.04.039
22. Elchalakani M, Zhao X-L, Grzebieta R (2002) Tests on concrete filled double-skin (CHS outer and SHS inner) composite short columns under axial compression. Thin-Walled Struct 40(5):415–441. https://doi.org/10.1016/S0263-8231(02)00009-5
23. Zhao X-L, Grzebieta R (2002) Strength and ductility of concrete filled double skin (SHS inner and SHS outer) tubes. Thin-Walled Struct 40(2):199–213. https://doi.org/10.1016/S0263-8231(01)00060-X
24. Han L-H, Tao Z, Huang H, Zhao X-L (2004) Concrete-filled double skin (SHS outer and CHS inner) steel tubular beam-columns. Thin-Walled Struct 42(9):1329–1355. https://doi.org/10.1016/j.tws.2004.03.017
25. Hassanein MFF, Elchalakani M, Karrech A, Patel VII, Daher E (2018) Finite element modelling of concrete-filled double-skin short compression members with CHS outer and SHS inner tubes. Mar Struct 61:85–99. http://doi.org/10.1016/j.marstruc.2018.05.002
26. Ayough P, Ramli Sulong NHH, Ibrahim Z, Hsiao P-CC (2020) Nonlinear analysis of square concrete-filled double-skin steel tubular columns under axial compression. Eng Struct 216:110678. http://doi.org/10.1016/j.engstruct.2020.110678
27. Hassanein MF, Kharoob OF, Gardner L (2015) Behaviour and design of square concrete-filled double skin tubular columns with inner circular tubes. Eng Struct 100:410–424. http://doi.org/10.1016/j.engstruct.2015.06.022
28. Wang F, Han L, Li W (2018) Analytical behavior of CFDST stub columns with external stainless steel tubes under axial compression. Thin-Walled Struct 127:756–768. http://doi.org/10.1016/j.tws.2018.02.021
29. Hassanein MFF, Elchalakani M, Karrech A, Patel VII, Yang B (2018) Behaviour of concrete-filled double-skin short columns under compression through finite element modelling: SHS outer and SHS inner tubes. Structures 14:358–375. http://doi.org/10.1016/j.istruc.2018.04.006
30. Garg A, Wani I, Zhu H, Kushvaha V (2022) Exploring efficiency of biochar in enhancing water retention in soils with varying grain size distributions using ANN technique. Acta Geotech 17(4):1315–1326. https://doi.org/10.1007/s11440-021-01411-6
31. Kushvaha V, Kumar SA, Madhushri P, Sharma A (2020) Artificial neural network technique to predict dynamic fracture of particulate composite. J Compos Mater 54(22):3099–3108. https://doi.org/10.1177/0021998320911418
32. Sharma A, Anand Kumar S, Kushvaha V (2020) Effect of aspect ratio on dynamic fracture toughness of particulate polymer composite using artificial neural network. Eng Fract Mech 228:106907. http://doi.org/10.1016/j.engfracmech.2020.106907
33. Sharma A, Kushvaha V (2020) Predictive modelling of fracture behaviour in silica-filled polymer composite subjected to impact with varying loading rates using artificial neural network. Eng Fract Mech 239:107328. http://doi.org/10.1016/j.engfracmech.2020.107328
34. Sharma A, Mukhopadhyay T, Rangappa SM, Siengchin S, Kushvaha V (2022) Advances in computational intelligence of polymer composite materials: machine learning assisted modeling, analysis and design. Arch Comput Methods Eng. https://doi.org/10.1007/s11831-021-09700-9

35. Sofi FA, Steelman JS (2021) Using committees of artificial neural networks with finite element modeling for steel girder bridge load rating estimation. Structures 33:533–553. https://doi.org/10.1016/j.istruc.2021.04.056
36. Cascardi A, Micelli F, Aiello MA (2017) An artificial neural networks model for the prediction of the compressive strength of FRP-confined concrete circular columns. Eng Struct 140:199–208. http://doi.org/10.1016/j.engstruct.2017.02.047
37. Tran V-L, Thai D-K, Kim S-E (2019) Application of ANN in predicting ACC of SCFST column. Compos Struct 228:111332. http://doi.org/10.1016/j.compstruct.2019.111332
38. Tran V-L, Thai D-K, Nguyen D-D (2020) Practical artificial neural network tool for predicting the axial compression capacity of circular concrete-filled steel tube columns with ultra-high-strength concrete. Thin-Walled Struct 151:106720. http://doi.org/10.1016/j.tws.2020.106720
39. Tran VL, Thai DK, Kim SE (2019) A new empirical formula for prediction of the axial compression capacity of CCFT columns. Steel Compos Struct 33(2):181–194. http://doi.org/10.12989/scs.2019.33.2.181
40. Zarringol M, Thai H-T, Thai S, Patel V (2020) Application of ANN to the design of CFST columns. Structures 28:2203–2220. http://doi.org/10.1016/j.istruc.2020.10.048
41. Tran V-L, Kim S-E (2020) Efficiency of three advanced data-driven models for predicting axial compression capacity of CFDST columns. Thin-Walled Struct 152:106744. http://doi.org/10.1016/j.tws.2020.106744
42. ANSYS Inc (2019) ANSYS 19.0 [Computer software], Canonsburg, PA
43. Sofi FA (2017) Structural system-based evaluation of steel girder highway bridges and artificial neural network (ANN) implementation for bridge asset management. Doctoral Dissertation, University of Nebraska-Lincoln, Lincoln, NE
44. ANSYS Inc (2019) ANSYS parametric design language guide. Release 2019, Canonsburg, PA
45. Sofi FA, Steelman JS (2017) Parametric influence of bearing restraint on nonlinear flexural behavior and ultimate capacity of steel girder bridges. J Bridg Eng 22(7):04017033. http://doi.org/10.1061/(ASCE)BE.1943-5592.0001065
46. Sofi FA, Steelman JS (2019) Nonlinear flexural distribution behavior and ultimate system capacity of skewed steel girder bridges. Eng Struct 197:109392. http://doi.org/10.1016/j.engstruct.2019.109392
47. Bažant ZP, Gambarova PG (1984) Crack shear in concrete: crack band microplane model. J Struct Eng 110(9):2015–2035. https://doi.org/10.1061/(ASCE)0733-9445(1984)110:9(2015)
48. Von Mises R (1913) Mechanik der festen Körper im plastisch-deformablen Zustand. Math Klasse 582–592 (in German)
49. Han L, Huo J (2003) Concrete-filled hollow structural steel columns after exposure to ISO-834 fire standard. J Struct Eng 129(1):68–78. https://doi.org/10.1061/(ASCE)0733-9445(2003)129:1(68)
50. Saenz L (1964) Discussion of 'equation for the stress-strain curve of concrete' by Desayi and Krishnan. J Am Concr Inst 61:1229–1235
51. ACI Committee 318 (2019) Building code requirements for structural concrete and commentary. American Concrete Institute
52. Hu H, Schnobrich WC (1989) Constitutive modeling of concrete by using nonassociated plasticity. J Mater Civ Eng 1:199–216. https://doi.org/10.1061/(ASCE)0899-1561(1989)1:4(199)
53. Mander JB, Priestley MJN, Park R (1988) Theoretical stress-strain model for confined concrete. J Struct Eng 114(8):1804–1826. https://doi.org/10.1061/(ASCE)0733-9445(1988)114:8(1804)
54. Richart FE, Brandtzaeg A, Brown RL (1928) A study of the failure of concrete under combined compressive stresses. University of Illinois
55. Hu H-T, Huang C-S, Wu M-H, Wu Y-M (2003) Nonlinear analysis of axially loaded concrete-filled tube columns with confinement effect. J Struct Eng 129(10):1322–1329. https://doi.org/10.1061/(ASCE)0733-9445(2003)129:10(1322)
56. Giakoumelis G, Lam D (2004) Axial capacity of circular concrete-filled tube columns. J Constr Steel Res 60(7):1049–1068. https://doi.org/10.1016/j.jcsr.2003.10.001

57. Mursi M, Uy B (2003) Strength of concrete filled steel box columns incorporating interaction buckling. J Struct Eng 129(5):626–639. https://doi.org/10.1061/(ASCE)0733-9445(2003)129:5(626)
58. Liang QQ (2009) Performance-based analysis of concrete-filled steel tubular beam–columns, part I: theory and algorithms. J Constr Steel Res 65:363–372. https://doi.org/10.1016/j.jcsr.2008.03.007
59. Marquardt DW (1963) An algorithm for least-squares estimation of nonlinear parameters. J Soc Ind Appl Math 11(2):431–441. http://doi.org/10.1137/0111030
60. MacKay DJC (1992) Bayesian interpolation. Neural Comput 4(3):415–447. http://doi.org/10.1162/neco.1992.4.3.415
61. American Concrete Institute (1984) State-of-the-art report on high strength concrete. ACI J 81(4):364–411

The manufacturer's authorised representative in the EU is Springer Nature Customer Service Centre GmbH, Europaplatz 3, 69115 Heidelberg, Germany. If you have any concerns regarding our products, please contact ProductSafety@springernature.com

Printed and bound by CPI Group (UK) Ltd, Croydon, CR0 4YY
30/09/2026
02225244-0001